The publisher and the University of California Press Foundation gratefully acknowledge the generous support of the Ralph and Shirley Shapiro Endowment Fund in Environmental Studies.

Reimagining Sustainable Cities

Reimagining Sustainable Cities

Strategies for Designing Greener, Healthier, More Equitable Communities

Stephen M. Wheeler and
Christina D. Rosan

UNIVERSITY OF CALIFORNIA PRESS

University of California Press
Oakland, California

Library of Congress Cataloging-in-Publication Data

Names: Wheeler, Stephen M., 1957- author. | Rosan,
    Christina, author.
Title: Reimagining sustainable cities : strategies for
    designing greener, healthier, more equitable
    communities / Stephen M. Wheeler and Christina
    D. Rosan.
Description: Oakland, California : University of
    California Press, [2021] | Includes index.
Identifiers: LCCN 2021012102 (print) | LCCN 2021012103
    (ebook) | ISBN 9780520381216 (cloth) |
    ISBN 9780520381209 (epub)
Subjects: LCSH: City planning—Environmental aspects. |
    City planning—Social aspects. | Urban ecology
    (Sociology)
Classification: LCC HT166 .W93 2021 (print) |
    LCC HT166 (ebook) | DDC 307.1/216—dc23
LC record available at https://lccn.loc.gov/2021012102
LC ebook record available at https://lccn.loc
    .gov/2021012103

30 29 28 27 26 25 24 23 22 21
10 9 8 7 6 5 4 3 2 1

Together we are very powerful, and we have a seldom-told, seldom-remembered history of victories and transformations that can give us confidence that yes, we can change the world because we have many times before.

—Rebecca Solnit

We live in capitalism. Its power seems inescapable. So did the divine right of kings. Any human power can be resisted and changed by human beings.

—Ursula K. LeGuin

Contents

Illustrations

Introduction

Imagine a city where housing is affordable, where each home produces more energy than it uses, and where people from different class, race, and ethnic groups live nearby and enjoy each other's company. Bikes and pedestrians outnumber cars, the air is clean, and sounds of birds and children's voices can be heard. Green spaces are visible from every dwelling. Little is wasted or thrown away. Women, people of color, and LGBTQ individuals feel safe, respected, and empowered. Businesses make decisions based on their benefit to workers, the public, and the environment as well as their own bottom line. Leaders focus on long-term collective well-being, and everyone collaborates in planning the community's future and undoing the wrongs of the past.

Imagine cities and towns, in other words, that will be sustainable and equitable far into the future.[1]

This vision may seem impossible, a dream so far from today's reality that it is not even worth considering. But something like it will unfold eventually, if for no other reason than that if humanity is to continue on this planet long term it will have to figure out how to live in such ways. Business As Usual (BAU), as we know it from the late twentieth and early twenty-first centuries, is unsustainable. The only questions are how soon societies can move onto a better path, how much damage they will do in the meanwhile, and to what lengths political forces will go to resist change.

This book seeks to take sustainable city discussions to a new level by considering the steps needed to truly address the climate crisis, social inequality, racial injustice, dysfunctional democracy, unaffordable housing, and other contemporary challenges. Past sustainable city books have often focused on topics such as green buildings, renewable energy, bike and pedestrian planning, and compact land development strategies. However, we want to go beyond those to explore more fundamental structural changes. Our belief is that it is necessary to reimagine institutional, economic, and political structures—what we call social ecology—in order to make sustainable communities possible. This reimagining will be a creative process, meshing changes in physical form with changes in policy, codes, institutions, and power structures, hence our use of the term *designing* in the title.

Our audience for this book includes all those who care about the future, so that includes you and everyone you know. Each one of us has the capacity to demand change in our own lives and in the systems that structure our lives. We have tried to write in a way that will be accessible to readers at multiple levels—to academics and professionals who design and plan urban areas as well as to students who study them and people who live in, work in, and care about them. We include footnotes for readers who want to dig deeper, and anecdotes and examples to help illustrate problems and potential solutions. For readers who are in a hurry or who want to flip quickly to solutions, we have included a concise table of strategies in each chapter. We hope that this mix of approaches will prove useful to a wide audience. Our goal is to be constructive and empowering rather than depressing and paralyzing, and to challenge readers to reimagine and work toward a better world. But that means first thinking critically about the structures that have created our current economic, political, cultural, and social systems: systems that are deeply in need of reimagining.

. . .

The need for dramatic change is urgent. Problems such as global warming, pollution, social inequality, structural racism, political dysfunction, motor vehicle dependency, environmental and climate justice, and loss of community bonds between people confront us daily. These are not new problems. It has been more than fifty years since the first Earth Day demonstrations demanded more ecologically conscious societies in 1970. It has been about the same length of time since Donella Meadows and others pioneered the term *sustainable development* in their 1972 book *Limits to Growth*.[2] It has been more than thirty years since James Hansen first testified to the US Congress in 1988 about global warming, and since *Time* magazine ran a "Planet of the Year" cover in 1989 with the caption "Endangered Earth."[3] Public protests against social and racial injustice are older still. However, every police killing of a person of color is a reminder that these structural problems need far more attention. The Covid-19 pandemic, the ensuing economic dislocation, and the strength of right-wing populism worldwide make the unsustainability of current social ecology trends ever clearer.

Societies worldwide are stuck in terms of addressing most of these problems. Elections come and go, but the quantity of greenhouse gases (GHGs) humanity pours into the atmosphere continues to grow almost every year.[4] Income inequality has continued to worsen in most nations.[5] The planet continues to lose up to one hundred thousand species annually because of human actions.[6] Plastics and toxics continue to spread through ecosystems. Oceans are increasingly polluted, coral reefs are bleaching, and seawater has absorbed so much carbon dioxide that it is becoming more acidic,[7] a trend that undermines marine food chains. Yet societies continue to operate as though the world will survive with minor policy changes instead of dramatic action.

Most worrisome, current societies seem unable to respond to the challenge. In particular, the constellation of political beliefs known as neoliberalism has hindered progress for some forty years now. This mindset asserts that the public sector (government) should be kept as small as possible, that voluntary action by the private sector and individuals can deal with collective problems, and that free markets exist and are the best way to allocate resources. Equally important, neoliberalism denies the centrality of power struggles based on class, race, and gender within societies and the ways that powerful interests warp social, political, and economic institutions and beliefs for their own benefit. It denies the deep historical roots of injustice and the need to proactively address them. It denies the need for a strong public sector to meet common needs. The neoliberal mindset has prevented effective social change and continues to blind many if not most of the world's leaders. The situation is increasingly dire.

Already the alarm bells are going off in response to current crises. In his book *Falter,* Bill McKibben argues that given global warming and other challenges, "The human experiment is now in question."[8] Luckily, people around the world are beginning to mobilize in response. Young people like Greta Thunberg and millions of others around the world are demanding action on the climate emergency. The slogan "I Can't Breathe" and the Black Lives Matter movement symbolize widespread desire to exorcise the legacies of colonialism, slavery, and systematic racism from our institutions and political structures. In 2020, millions of Americans waited in long, socially distanced lines risking their lives at the polls to support democracy. A collective process of rethinking and reimagining societies has begun.

Neoliberalism is facing increased criticism worldwide, but it is not yet clear what will take its place. Authoritarian populism is one possibility, represented in recent years by leaders such as Donald Trump in the US, Boris Johnson in the UK, Jair Bolsonaro in Brazil, Recep Erdogan in Turkey, Viktor Orbán in Hungary, Rodrigo Duterte in the Philippines, and Narendra Modi in India, as well as more traditional authoritarians such as Vladimir Putin in Russia and Xi Jinping in China. The even stronger merger of business, military, and political interests known as fascism is a possibility as well. These approaches threaten basic human rights and democratic principles as well as the health of planetary ecosystems. Positive alternatives are much needed. We see planning for sustainable cities as a critical part of an alternative politics that can bring about a more just and sustainable world.

. . .

The term *sustainable development* refers to efforts to ensure long-term human and ecological welfare. Ever since 1972 the discourse around it has served as a leading way to conceptualize progressive social change. Three main themes within this discourse are the need to think long term, the need to think holistically and intersectionally about both problems and solutions, and the need to be proactive so as to solve these challenges.

Some argue that *sustainable development* is an oxymoron, equating "development" with destructive, overly consumptive ways of living. Others view the term as static, connoting some impossibly idealized steady state of society. However, in our view sustainability does not mean either continuing the status quo or aiming at a static utopia. Rather, it connotes a process of continually and actively moving in directions that promote ecological health, social equity, quality of life, cooperation, and compassion.

The urban sustainability agenda has evolved greatly since the planning and design professions began to embrace this concept in the 1990s. Planning for the climate crisis has become vastly more urgent, and new emphases on urban food systems, safe and affordable water, structural racism, public health, and reforming capitalism and democracy have emerged (see figure 1). Although the sustainability concept is often

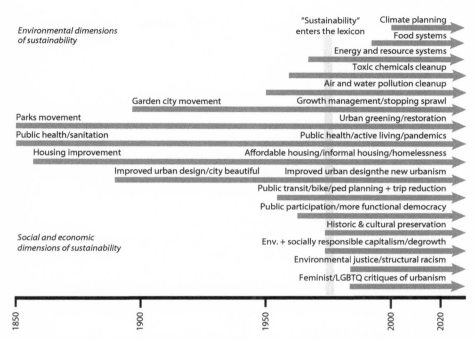

FIGURE 1. Sustainable city discussions have evolved greatly over the past forty-plus years. Recent movements emphasize climate action planning, environmental justice, urban food systems, public health, economic degrowth, and structural racism. The need for structural change in societies has become ever more apparent. (S. Wheeler.)

co-opted to refer exclusively to environmental dimensions of change, any meaningful discussion of it must include goals of social and racial equity and economic transformation, both within industrialized countries and between the Global South and the Global North. Sustainability discussions must also address the structural barriers and injustices that to date have prevented the necessary level of action.

For each sustainable community challenge there is, of course, no perfect solution. But in many cases, cities somewhere in the world have pioneered strategies that can make a difference. We seek to outline these approaches in each chapter, with the hope that political discourse can be widened to more fully consider them. Best strategies for any given community or society will depend on the context and may mix together multiple approaches that have been tried elsewhere. But the first step is usually to acknowledge the need for change and then to identify possible solutions—in other words, to reimagine.

Any sustainability challenge can be investigated from multiple angles. For example, some may feel that the root cause of a problem such as global warming is capitalism (a large-scale economic system). However, others might argue that socialist countries have had high GHG emissions too. Still others could assert that global warming is caused by deficient government institutions (poorly functioning democracy, bad laws, and/or weak enforcement) or misplaced market incentives (too low a price on carbon). Further observers might argue that the root problem is one of values, ethics, structural racism, and/or a lack of moral responsibility.

In this way every sustainability problem can be seen to have multiple roots, and multiple strategies are needed to address it. Global warming (or the climate emergency), for example, will likely require strong regulatory and market strategies *and* fundamental revisions to capitalism *and* changes in values and lifestyles *and* improvements in democracy so that better societal decision-making can come about. These represent strategic intersectional changes throughout the social ecology.

On the bright side, humanity has in fact made progress on some sustainability issues. Many societies have cleaned up air and water, improved their energy efficiency, and instituted modest levels of resource recycling. Hunger has fallen worldwide,[9] and global population looks on track to stabilize this century, albeit at a very high level (close to eleven billion people).[10] Many countries do better at protecting the rights of women, the differently abled, LGBTQ individuals, and people of color than they did in 1970. However, lots more remains to be done, and with rapid population growth, the climate emergency, and technological development the list of challenges keeps expanding.

It is particularly encouraging that in 2015 the world's nations acting under United Nations auspices endorsed seventeen Sustainable Development Goals (SDGs) for the year 2030. These goals establish an excellent road map for sustainability. They include eliminating poverty and hunger (SDGs 1 and 2), providing quality primary and secondary education and early childhood development for all (SDG 4), achieving full equality for women and girls (SDG 5), reducing inequalities (SDG 10), and taking action on climate (SDG 14).[11] Many SDGs could be refined and expanded, but the SDGs show that as a species we have increasing agreement on what needs to be done. Now we have to act.

World leaders agreed to an even more specific Urban Development Agenda in 2016.[12] This sets forth a vision of "just, safe, healthy, accessible, affordable, resilient and sustainable cities and human settlements." It also articulates a human right to adequate housing and specific goals

FIGURE 2. Adopted in 2015 for the year 2030, the UN Sustainable Development Goals establish a strong, holistic foundation for thinking about sustainable societies. Activists and organizations worldwide are working on ways to implement them. (UN News, "Sustainable Development Goals Kick Off with Start of New Year," December 30, 2015, www.un.org/sustainabledevelopment /blog/2015/12/sustainable-development-goals-kick-off-with-start-of-new-year/.)

of civic engagement; safe, inclusive, accessible, and green public spaces; an embrace of diversity; strengthened social cohesion; resource-efficient economies; and decent work for all (see figure 2). Although these UN formulations are far from perfect, humanity now has agreed on some common goals for sustainable societies and sustainable cities.

. . .

In this book, our theoretical approach is that of social ecology. This perspective analyzes how strategic, interrelated changes on many dimensions—institutions, economics, politics, values, behaviors, technologies, and cognition—can help societies evolve in healthier and more just directions.[13] The concept has roots in nineteenth-century thinkers including Darwin, Marx, and Spencer,[14] and it continues through the Chicago School of urban sociology beginning in the 1910s, the work of anarchist social philosopher Murray Bookchin in the 1960s and 1970s, and more recent writers including Meadows, Stewart Brand, Richard Norgaard, and Elinor Ostrom.[15] It also draws on perspectives from Buddhism (which strongly emphasizes holistic awareness) and many other religious and indigenous traditions worldwide that connect humanity with nature and develop frameworks of ethical responsibility.[16] Social ecology is the natural extension of holistic, systems-oriented thinking, integral to ecological science and many of the world's wisdom traditions. These days the social ecology challenge very much includes rethinking democracy, which, to put it mildly, is not functioning well in many parts of the world. Societies need to take collective responsibility and decision-making to the next level and rethink social, political, and economic systems so that all people can exist long term on a small and changing planet.

The social ecology worldview is quite different from the dominant twentieth-century mindset of modernity, which emphasized reductionistic, linear thinking, and specialization of knowledge. Modernity placed great faith in quantitative research and technological solutions at the expense of more qualitative analyses of power, cognition, and values within societies. Highly specialized disciplines emerged in both academic and professional worlds with their own associations, conferences, journals, and textbooks. People operated in silos. Scientists developed knowledge but didn't necessarily link it to action, situate it in historical context, or question the assumptions that underpinned it. Governmental and academic departments focused on pieces of the puzzle but not systems as a whole. Attempts at being interdisciplinary often meant that people briefly emerged from their disciplinary silos but then soon returned to them. It is increasingly clear that this BAU approach will not help address many of the toughest sustainability challenges.

The dominant worldview of industrial societies is slowly changing. People everywhere are learning to think more holistically.[17] They are also learning to incorporate diverse viewpoints that are more representative of the world's diverse population. Words like *intersectionality* are popular these days among those trying to analyze complex situations from multiple points of view. Increasingly people are being not just interdisciplinary but transdisciplinary—focusing their work on big-picture challenges such as global warming (or weirding), healthy food systems, and environmental justice and collaborating across disciplines and scales to get things done. This type of thinking can be messy and difficult. It may require us to question our own biases, positionality, and blind spots. However, it is essential in order to understand problems, reimagine our communities, and identify equitable paths forward.

Thinking holistically about social ecologies means not just thinking across disciplines but thinking across scales. Big-picture changes at national or global levels can help address problems, but local and regional action is essential as well. Bumper stickers in the 1970s used to tell people to "think globally and act locally." But the need is to think at multiple scales and act at whatever scales we can, from our personal homes and lifestyle choices to the larger systems encountered in our professional and community lives.

We focus in this book on the underlying structural problems and social ecology solutions that are most important for moving toward sustainable and equitable communities. New technologies are useful but not enough. Electric cars, for example, are an important way to help reduce GHG emissions. However, these vehicles will be part of a carbon-neutral future only if the electricity they use is generated renewably, embodied emissions in the vehicles themselves are minimized, and the vehicles are affordable to a broad and diverse range of people. Moreover, heavy use of motor vehicles even if electric will still aggravate urban sustainability problems such as traffic congestion, pedestrian safety, suburban sprawl, and the cost of maintaining far-flung infrastructure. So at best this technology is only a partial solution. Reducing our dependency on private motor vehicles of any type is essential as well. This requires further changes in lifestyle, economics, infrastructure, and urban design.

Indeed, an excessive focus on technology often distracts people from much-needed changes in institutions, values, behaviors, investments, and worldviews. Societies have had most of the knowledge and technology needed for sustainability for generations. Most cities were far more sustainable seventy or one hundred years ago when their residents walked or rode streetcars rather than driving, returned glass bottles to be washed and reused, ate food coming from farms just outside the city, and bought virtually nothing made of plastic. This is not to propose a return to the socially unjust world of 1900 or 1950, with its racism, sexism, and homophobia, but to point out that technology by itself does not solve problems and can often lead to new problems if we do not think intersectionally. The fundamental obstacles to sustainability are political, institutional, economic, social, and behavioral rather than technical.

In particular, decision makers the world over have not focused enough on social equity and racial justice as a critical part of social ecology solutions. Poor people have the least power within most societies. They are unseen, unheard, and often mistreated and oppressed by industry, government, and elites. People of color and immigrants are particularly marginalized and exploited. Green, beautiful neighborhoods where only wealthy, white citizens live are not sustainable. Neither are urban regions in which some neighborhoods are poor and disinvested while others have good schools, clean air, and safe drinking water. Finally, it is wrong for some parts of the world to be underdeveloped, with billions of people living without sanitation, housing, food, health care, and basic human rights, while in other parts people over-consume, pollute, and are responsible for large quantities of GHGs. All of us need to work together to undo these inequalities.

. . .

We focus on cities because the world is increasingly urban. In 1950 only 30 percent of the world's population lived in urban areas. By 2000 this figure had grown to 47 percent, in 2020 it was 56 percent, and by 2050 the city-dwelling proportion of humanity is expected to reach 68 per-cent.[18] Like it or not, our species is urbanizing. Making relatively dense, human-created places sustainable is one of the twenty-first century's foremost challenges and opportunities.

However, in "cities," we include suburbs, exurbs, and entire metro-politan regions, since these are interconnected. Regional coordination between these types and scales of communities is essential to address sustainability problems. Since such coordination has not happened in many parts of the world to date, higher levels of government will often need to step in to make it happen by providing a policy environment and the funding that supports local creativity and is conducive to leveraging cooperation across municipalities. We will return to this theme in later chapters.

Given the risk of pandemics like Covid-19, there has been a recent rethinking of the role of cities. A world where social distancing is neces-sary at times brings public transportation and crowded public spaces into question. Some argue that the threat of pandemics means the end of cities. Newspapers have run stories about urbanites retreating to the suburbs or the country. However, such threats are also an opportunity to imagine better cities. Rather than abandoning the urban experiment, we argue that Covid-19, the movement for racial and social justice, and the climate crisis all should bring cities to the forefront of public attention. This is an oppor-tunity to create cities that are resilient to such challenges, while continuing to make the most of inherent urban advantages of reducing everyone's ecological footprint and promoting culture and diversity.

Cities need to take sustainability action quickly, since problems are often most severe within their bounds. According to C40 Cities, a con-sortium of municipalities working to meet the Paris Accord climate change goals, 70 percent of large urban regions have already felt the impacts of climate change, and 90 percent are coastal and thus extremely vulnerable to sea-level rise, more intense storms, and other climate-related dislocations.[19] Think about that: 90 percent. Urban areas also produce 70 percent of GHG emissions, so the process of moving to climate neutrality will need to start there.[20] Cities witness some of the starkest social inequalities, the worst pollution, the most pressing environmental justice concerns, and the most extreme waste of resources. New York City, for example, generates in excess of fourteen million tons of waste annually and spends $2.3 billion to dispose of it as far away as China.[21]

However, urban regions are also the laboratories in which we can demand more and expect better. Since antiquity, cities have been places of experimentation and creativity. They tend to be more diverse than societies as a whole and more tolerant. Urban regions often have more progressive politics than rural areas, and their character and economies of scale allow them to produce more sustainable outcomes. Cities and towns across the globe are already promoting more sustainable forms of transportation, energy use, building construction, food systems, and materials processing. They are planning and building for a sustainable and equitable future that will look very different from today.

. . .

How do we start reimagining sustainable cities? Volumes could be written about any of the questions addressed in this book. As authors we can suggest the needed range of strategies to address each urban sustainability problem but cannot exhaustively delve into every topic. We do not claim to be unique authorities on any of these subjects. Our main contributions are to ask questions that expand the range of debate and to suggest the outlines of a more sustainable future.

Big-picture thinking about so many subjects is difficult, and we apologize in advance for any errors or misplaced emphases. But we feel this type of holistic exploration is very much needed and can be a crucial resource for students at many levels and for interested citizens and professionals. We encourage readers to use each chapter as a stepping-stone to their own further explorations.

We live in the United States and that is our primary frame of reference. But the US exists in a global context. All cities and societies everywhere need to learn from each other. A sustainability approach requires each of us to think about where our clothes were made, our cars are built, and our food is grown, and how our mental constructs are generated. It also requires us to ask, Are we okay with a global system that has left some countries underdeveloped while others overconsume and Elon Musk rockets cherry-red Teslas into outer space? For such reasons, we try to take a global approach as much as possible. We include international examples and seek to spell out implications of various strategies for many types of communities worldwide.

We likewise seek to strike a balance between focusing on urban areas and discussing society-wide change. Cities can do many things themselves, but they are also dependent on action at other levels. To arrive at sustainable cities, a framework of change is needed that includes steps by many different players across scales. Our aim is to spell out this overall vision rather than to focus on the limited subset of actions that can be taken at the municipal scale through the door that says "Planning Department." We see planning as a collective responsibility.

We are optimistic about the future. Generational shifts are under way in many countries. Generation Z is the most diverse and politically engaged yet. Its members have grown up knowing that their success and even survival depend on their willingness to address issues such as global warming and inequality. Intuitively, many young people know that BAU must change.

We hope this mix of urban sustainability questions, context discussion, strategies, and examples will prove useful. Beyond that, we hope it will inspire readers to take action in their own lives and communities. There is always some visionary local action that one can pursue to generate a small-scale example of change. Each of us can look for creative action within our own lives, homes, and communities to reimagine a more sustainable world, in turn inspiring others with commitment, courage, and action.

How Do We Get to Climate Neutrality?

In 2018 Africa experienced its highest temperature ever, when Ouargla, Algeria, experienced a high of 124.3°F (51.3°C).[1] That year a nuclear reactor in Sweden had to be shut down because the local seawater had become too warm to cool it.[2] Grocery stores had to be kept open around the clock as emergency cooling centers, since most Swedish homes don't have air conditioning, Sweden not being historically a hot country.

In 2019 the city of Churu in India reached a temperature of 123.4°F (50.8°C), a few tenths of a degree short of the all-time Indian record set only three years earlier. Other North Indian cities baked as well, and the heat wave lasted for thirty-two days, the second-longest ever recorded. At least thirty-six people died.[3]

The next year the remote Siberian town of Verkhoyansk recorded a temperature of 100.4°F (38°C), the highest temperature ever recorded north of the Arctic Circle.[4] That year a heat wave in California produced a new world record temperature: 130°F (54°C) in Death Valley.[5] The same heat wave produced electrical storms that sparked fires across Northern California, one of them generating intense winds that produced a phenomenon new to many: the "firenado."

The onward march of heat records is only one manifestation of the climate crisis.[6] Drought and wildfires in 2010 destroyed much of the Russian wheat crop, leading global wheat prices to rise 84 percent. Supertyphoon Haiyan in 2013, with sustained winds of more than 195 mph, killed at least 6,700 people in the Philippines and left hundreds of thousands homeless. The 2018 California wildfires burned 1,893,913 acres, obliterated the town of Paradise, and led to insurance claims of more than $12 billion. The unprecedented extent of those fires was a shock until in 2020 a new round of fires burned 4,197,628 acres.

We start our reimagining of sustainable cities with the climate crisis, since this is the largest current sustainability challenge and the costs of inaction are enormous. Moreover, according to the editors of *The Lancet,* "Both Covid-19 and the climate crisis have exposed the fact that the poorest and most marginalised people in society, such as migrants and refugee populations, are always the most vulnerable to shocks."[7]

Climate is also key to many other sustainability issues, illustrating ways that these challenges and their solutions are deeply intertwined. The goal is often seen as climate neutrality, in that the net quantity of greenhouse gases (GHGs) humanity contributes to the atmosphere must be zero, or preferably negative as societies figure out better ways to sequester the carbon that they have already emitted in soils, in forests, or perhaps underground. However, for particular communities, companies, or institutions *climate neutrality* can be a controversial term if it includes purchase of offsets that allow continued emissions (often affecting disadvantaged communities) on the promise that GHGs will be reduced elsewhere. In such cases, "carbon free," "fossil free," or "complete decarbonization" may be preferable goals ensuring that these entities do their part to rapidly end GHG emissions and achieve environmental justice goals.

Moving toward climate neutrality means actions at many scales— local, regional, state/provincial, national, and international. Addressing the climate crisis will probably also require more effective governance at each level, in turn creating the capacity to address a host of environmental, economic, and social needs. Conversely, *not* addressing the climate problem will compound sustainability challenges at many scales and create massive environmental justice problems as vulnerable populations suffer increased poverty, hunger, displacement, and violence.

So let us imagine communities that have quickly and decisively taken action to eliminate fossil fuels from buildings, industries, and vehicles. They generate 100 percent of their energy from renewable sources. They have greatly reduced methane emissions related to landfills and people's diets. They have eliminated emissions of minor GHGs and spearheaded programs to sequester atmospheric carbon in forests and soils. And they have taken these steps in ways that are equitable and improve quality of life for everyone.

How can such a future come about?

In this chapter we'll consider the evolution of the climate crisis and eight main strategies to end humanity's increase of atmospheric GHGs. We'll also discuss how the social ecology around this issue might be changed. The latter is particularly important because, as with other topics, the question of how to get around structural obstacles that prevent climate action can't be separated from the question of how to reduce GHG emissions themselves. System change is needed, and this must go far deeper than many of the policies usually considered.

THE EVOLUTION OF A CRISIS

Concern about global warming dates back to 1898, when Swedish scientist Svante Arrhenius first calculated how much the Earth would be likely to warm with a doubling of atmospheric CO2, likely by 2100 on humanity's current trajectory. Arrhenius estimated 4°C or 7.2°F, an amazingly accurate projection given that he did all of his calculations by hand. Recent estimates give a range of between 2.6° and 4.1°C, or 4.7° and 7.4°F.[8]

During the first half of the twentieth century the possibility of global warming seemed far away. Many scientists believed the oceans would absorb all of the carbon dioxide we were producing by burning fossil fuels. However, this situation changed in the 1950s. American scientists Hans Suess and Roger Revelle determined that the oceans would not in fact be able to solve the problem by completely soaking up CO2, and they testified to the US Congress in 1957 that radical climate changes might occur. Geochemist Charles Keeling began taking continuous measurements of atmospheric carbon dioxide at Mauna Loa Observatory in Hawaii in 1958, producing hard evidence that concentrations were rising. Yet despite front-page media coverage in the late 1950s and authoritative research reports in the 1960s and 1970s, the world's governments resisted action for another thirty years.

By the late 1980s the problem could no longer be ignored. The United Nations Environment Program and the World Meteorological Organization established the Intergovernmental Panel on Climate Change (IPCC) in 1988 to coordinate scientific research, even though some environmentalists at the time argued that this body was not needed since the science was already clear enough to justify action.[9] That body has released assessment reports every five years summarizing consensus science on the issue. In 1992 the world's nations agreed to the United Nations Framework Convention on Climate Change (UNFCCC), acknowledging the climate crisis and beginning a long series of international meetings to produce binding treaties reducing emissions. The first of these was the 1997 Kyoto Protocol, agreed to by more than 190 countries, which established GHG reduction targets for thirty-five industrialized nations by 2008–12 averaging around 7 percent below 1990 levels. Exact mechanisms for reductions were left up to each country. Developing nations such as China and India weren't covered by the agreement, partly because of beliefs that developed countries should bear greater responsibility for their historic emissions and had greater capacity to reduce emissions.[10]

A growing tide of neoliberal politics in the 1990s and 2000s undercut actions toward meeting the Kyoto goals. A disinformation campaign by the fossil fuel industry helped as well. Exxon and other companies knew the potential global warming effects of CO_2 emissions as early as the 1950s but consciously decided in the 1980s to manipulate the media so as to confuse the public, putting forth false allegations about "unsettled science."[11] Opportunistic politicians on the right of the political spectrum worked hand in glove with the fossil fuel industry. The George W. Bush administration withdrew the US from the Kyoto process in 2001, and Canada left in 2011 as tar sands development became a priority for conservative prime minister Stephen Harper's government. By 2012 fewer than half of the thirty-five countries had met their Kyoto targets, and most of those were eastern European nations whose emissions had decreased because of economic decline. Even if goals had been achieved, the small Kyoto GHG reductions were not nearly enough to put the world on a path toward climate neutrality.

Nations at the 2015 Paris conference of the UNFCCC negotiated a new framework (the "Paris Agreement") asking each country to develop its own voluntary targets and programs to hold global warming to 1.5° C. However, in the following years the Paris Agreement made no significant difference in the world's emissions, which continued to increase except for a small dip due to the Covid-19 pandemic and recession.

This bleak picture in terms of international negotiations is not the only story, however. Many national governments have now acknowledged the necessary policy goal—very low or zero emissions by midcentury—even though few countries have met their own targets to date. Equally importantly, many states, provinces, and municipalities have taken leadership in adopting strong GHG mitigation policies. Examples range from Vancouver, British Columbia, reducing its residents' driving by 36 percent per capita between 2007 and 2017 through improved public transit and bike

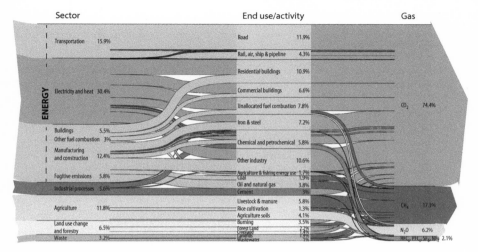

FIGURE 3. Strategies for carbon neutrality will need to cross virtually every economic sector and end use. (World Resources Institute, "World Greenhouse Gas Emissions: 2016," www.wri.org /data/world-greenhouse-gas-emissions-2016.)

facilities to Dubai's development, in 2013, of the largest solar park in the world, a facility expected to have a capacity of five gigawatts by 2030, as much as five large nuclear power plants.[12] Such actions are very real steps toward climate-neutral cities and towns.

STRATEGIES FOR CLIMATE NEUTRALITY

Figure 3 shows that transportation, buildings, the production of goods and services, agriculture, and land use change produce most GHG emissions worldwide. Strategies for climate neutrality must zero out these basic sources. Initiatives to do this can be grouped in eight broad categories, with a ninth, geoengineering, to be considered as a last resort if all else fails (table 1).

Require Carbon-Free Electricity

Since production of electricity is responsible for a large share of GHG emissions almost everywhere, replacing coal, oil, and natural gas as electricity-generating fuels with renewable energy is an essential starting point for climate neutrality. The public sector can then require that electricity be used instead of fossil fuels for vehicles, factories, and home

TABLE 1   STRATEGIES FOR CLIMATE NEUTRALITY

Strategy	Description
Require carbon-free electricity.	Use renewable portfolio standards or other means to ensure 100% carbon-free electricity as the cornerstone of a carbon-neutral economy.
Require zero-net-energy or plus-energy buildings.	Reduce building and appliance energy consumption to the lowest level possible; require buildings to be all-electric; generate electricity on-site; develop programs to retrofit existing buildings.
Reduce motor vehicle use and electrify vehicles.	Reduce motor vehicle travel through better urban design, improved alternative modes, pricing and informational strategies, and lifestyle changes. Convert all vehicles to electricity so as to use no fossil fuels.
Limit air travel.	Adopt pricing or rationing strategies to limit air travel as long as no alternative technology exists.
Place a rising price on carbon.	Add a powerful price disincentive for fossil fuel use, rising annually to $100/ton of CO2e or more, to bring about change throughout the economy and in individual behavior. Social equity impacts would need to be offset.
Adopt climate-friendly lifestyles.	Reduce personal consumption of carbon-intensive products.
Move toward climate-friendly agriculture and food systems.	Change diets to deemphasize meat, dairy, and processed foods; make other food system improvements including reduced waste and climate-friendly agricultural practices.
Sequester carbon in ecosystems.	Since some emissions will be inevitable, take carbon out of the atmosphere by sequestering it in forests and rangelands.
Consider geoengineering as a last resort.	If absolutely necessary, consider last-ditch interventions such as injecting aerosols into the stratosphere, seeding oceans with iron, or placing sun-shades in space to slow global warming while humanity implements other changes.

heating and cooling—indeed, almost all human activities except air travel. The goal is to bring about all-electric communities powered by renewably generated electricity.

Photovoltaic panels, large-scale concentrated solar power systems, and wind turbines can now produce electricity more cheaply than any fossil fuel in most parts of the world.[13] If national and state governments eliminate fossil fuel subsidies and price in the externalities associated with fossil fuel use, their use will expand ever more rapidly. Geothermal plants, small-scale hydropower, tidal power, and biomass are other potentially cost-effective sources of clean electricity. To further speed the transition, governments could adopt modest incentives such as tax credits, low-interest financing, and fast-tracked permitting for both small- and large-scale renewable energy systems.

One successful tool used to date has been for the public sector to adopt "renewable portfolio standards" (RPSs) mandating that utilities procure a certain percentage of their power from renewable sources. Although such policies are usually set at a national or state scale, some cities own or oversee their own utilities and can establish such standards. Thirty-eight US states have adopted RPS requirements, which have led to about half the growth in US renewable energy production since 2000.[14] California has adopted an RPS goal of 100 percent clean electricity by 2045; New York's is 70 percent by 2030.[15] Countries worldwide are also using this strategy. The European Union has a goal of 32 percent renewable electricity by 2030,[16] as well as overall climate neutrality by 2050. Germany actually produced 42 percent of its electricity from renewables in 2019 and has set a minimum goal of 80 percent by 2050.[17]

The need everywhere is to establish a requirement of 100 percent renewable electricity as soon as possible. One usual counterargument is that because wind and solar are intermittent, fossil fuel backup is needed to provide 24/7 supply and stabilize the grid. However, various technologies—including batteries and pumped-storage hydroelectricity—have the potential to store renewably generated electricity for backup purposes. Eventually every building might be built with photovoltaic panels covering the roof and its own internal battery storage so that electricity generated during the daytime could be stored for evening use, thus leveling out demands on the utility grid.

One main institutional problem is that private electric utilities, in those countries which have them, are often reluctant to move rapidly toward green power, especially if generated by households and other small-scale local producers. Reasons are understandable: utilities have sunk costs in large fossil fuel plants; they profit the most from large-scale centralized energy systems that they control; and they simply have no incentive to change their practices. However, the problem remains a serious one. Their large size and political influence often allow utilities to control regulatory agencies and avoid tough GHG regulation. To get around this problem, the public sector, including cities, may need to take over private utility companies and run them for the ultimate benefit of the public and the environment rather than shareholders.

"Public power" has long been a rallying cry historically, and American urban areas such as Sacramento, California, Boulder, Colorado, and Long Island, New York, have municipalized their utilities in the past.[18] Nonprofit electric cooperatives dominated from the beginning in many rural areas, since markets were too small for private utilities to be interested. More than two thousand American communities with forty-nine million people have nonprofit, community-owned, and locally controlled utilities.[19] But that leaves almost three hundred million US residents served by for-profit utilities. Internationally the picture is much the same: many of the world's largest energy providers are private for-profit companies, including Enel in Italy, EDF and Engie in France, TEPCO in Japan, KEPCO in South Korea, and Iderdrola in Spain.[20] Since the electric utility sector is so vital to GHG reduction efforts, a worldwide expansion of the public power movement may be needed.

Indeed, public power is already spreading in the US through another mechanism: Community Choice Energy. Under this system, municipalities set up nonprofits to purchase renewably generated electricity for their residents and distribute this over the regional utility's grid. This innovation allows cities to bypass the utility and provide residents with clean (and cheap) power long before those slow-moving companies would have done so. Begun in the 1990s, this method now provides climate-friendly power to thirty million Americans in more than 1,500 cities and towns.[21]

Require Zero-Net-Energy or Plus-Energy Buildings

The largest share of GHG emissions in many communities comes from the energy used to heat and cool buildings and run appliances and machinery inside them. Here again proactive planning is necessary to move societies toward climate neutrality. In particular, cities, states, and countries can update building codes to require all new buildings to be all-electric (essentially banning fossil fuels such as oil and gas from within them) or else zero-net-energy (ZNE; generating as much energy as they consume) or plus-energy (generating more than they consume).

Such standards are not far off. In 2017, nineteen large cities, including Melbourne, New York, Tokyo, and Johannesburg, signed a pledge to make all new buildings carbon-neutral by 2030.[22] California's building code is already approaching this level, requiring very high levels of energy efficiency and photovoltaic panels on all new homes. A number of California cities have banned the use of natural gas in new structures.

Even in cloudy, cool Germany ZNE buildings are possible,[23] and a *Passivhaus* movement has long been underway, creating structures with no active heating and cooling systems.[24]

If buildings and appliances are highly energy-efficient, photovoltaics on the roof or exterior walls can usually come close to supplying enough electricity to run them. Where trees shade roofs, builders could be allowed to buy solar power from elsewhere. If grid electricity is 100 percent renewable, producing power on the building itself is less essential, though having buildings ZNE or close to it is still desirable to help create more resilient local energy systems and minimize energy demand. Cost is less of an issue with each passing year. One recent study found that near-ZNE structures are already feasible and cost-effective in fourteen cities across Europe with varying climates.[25] Another study by the Rocky Mountain Institute found that the cost of electrification could be significantly cheaper than gas.[26] Cities such as Berkeley, California, and Brookline, Massachusetts, were among the first in the US in the late 2010s to require all new buildings to be all-electric. Other cities can follow their lead.

To accelerate the transition to ZNE for *existing* buildings, local governments can require gradual upgrades. States such as California already require energy efficiency upgrades to ductwork and insulation when heating or air conditioning units are replaced. Cities such as Berkeley, Minneapolis, and Boise have adopted ordinances requiring owners to conduct energy assessments of homes at time of sale. The next step could be to require them to upgrade buildings. To help finance such upgrades, local governments can set up property assessed clean energy (PACE) programs, which are essentially loan funds that allow property owners to make energy improvements up front and pay back the loan over time through an additional assessment on the property tax bill. The public sector could also make tax credits available for energy retrofits and could prioritize and target assistance for low-income households.

However, if necessary various levels of government could simply subsidize ZNE building construction or retrofits outright. The $230 billion annual cost of the 2017 US tax cut for businesses and the wealthy, for example, could provide a subsidy of $50,000 apiece to make each of the approximately four hundred thousand new homes built in the country each year ZNE, while also subsidizing a $22,500 energy retrofit of each dwelling within the 140-million-unit existing housing stock over fifteen years.[27] The country would be making its entire housing stock nearly carbon-neutral rather than doling out public money to rich people. People at all income levels would then save billions on their utility bills and we would improve public health and reduce climate impacts.

A related strategy is for communities to require that fossil fuel–powered appliances be replaced with electric ones whenever they wear out. Furnaces, hot water heaters, and gas dryers typically need to be replaced every twenty to thirty years. Cities, states, or countries could simply require that building owners install electric heat pump–powered units instead of fossil fuel–powered devices. Heat pumps are already cost-competitive in many places.[28]

Energy efficiency regulations across the board could be far stronger. There is no reason, for example, why incandescent light bulbs should still be sold anywhere in the world given the availability and relatively cheap prices of LED and CFL versions that use one-fifth of the energy.[29] Brazil, Venezuela, Australia, China, and the European Union have long since adopted programs to phase out incandescent bulbs. The EU's ban is estimated to save fifteen million tons of GHG emissions annually.[30] However, the Trump administration in 2019 reversed US policy to reduce use of incandescents, and many countries in the Global South have yet to adopt bans.

Other standards for residential, commercial, and industrial energy uses could be strengthened. One of the last studies published by the Obama administration showed that office equipment could be 41 percent more energy-efficient using existing technology, and dishwashers, cooking appliances, water heaters, and laundry equipment 60 percent, 24 percent, 35 percent, and 50 percent more efficient respectively.[31] If increased cost is an issue for buyers, especially low-income individuals, the public sector could make subsidies available.

To fully embrace a climate neutrality goal, regulation needs to also take into account the embodied emissions within products and services. Historically this has been fairly difficult to do because it means tracing back the supply chain of components that went into a product and also calculating emissions resulting from disposal of the product. However, knowledge on this topic is developing rapidly. Good estimates now exist of embodied energy within many goods. One study of 866 consumer products found that on average 44 percent of emissions associated with a product occurred before the consumer acquired it, 23 percent during its use, and 32 percent following its disposal.[32] Regulation that comprehensively takes all of these emissions into account is urgently needed. Information about the embodied emissions in our buildings and possessions can help us make better decisions about whether to repair or retrofit them and help companies be more accountable.

Reduce Motor Vehicle Use and Electrify Vehicles

Since motor vehicles are one of the largest GHG sources worldwide and even if electric have substantial embodied emissions and other sustainability impacts, the path to climate neutrality includes reducing their use to the lowest level possible. This will also have many other sustainability benefits in terms of reduced traffic congestion, improved human safety, quieter cities, time savings, and improved air quality and public health.

Four main strategies can reduce driving: creating more compact, well-balanced, and well-designed communities so as to reduce people's need to travel; providing a better range of transportation alternatives, especially public transit, walking, and biking; improving price signals and information so as to discourage driving and encourage alternatives; and moving toward non–motor vehicle–oriented lifestyles and behavior. Chapter 6 addresses these approaches in greater detail.

Since motor vehicles will never entirely disappear, those that remain will need to become all-electric, thus ending this enormous use of fossil fuels. Britain and France have already announced that new gasoline and diesel vehicles cannot be sold in their countries after 2040. In 2020 California announced that it will ban the sale of new gas-powered cars and trucks by 2035, a step that is expected to reduce the state's GHG emissions by 35 percent.[33] India has announced a target of 2030 for such a transition, and Norway is aiming for 2025.[34] China has said that it will set such a deadline as well.

Electric vehicle technology is advancing rapidly. Millions of electric cars, trucks, buses, and trains are already in operation worldwide. Problems such as the limited range of battery-powered vehicles are fading. The current generation of electric cars has a range of around 250 miles before recharging, and the next generation will have a range of around 500 miles.[35] Eventually most vehicles will likely have solid-state batteries that can be recharged up to 80 percent in less than ten minutes, making the process similar to refueling gasoline-powered vehicles.[36]

Since cost is a major barrier for many households to shift to electric vehicles, a second step is to subsidize the transition for consumers. To make them more affordable, Norway initially waived its high vehicle import duties as well as registration and sales taxes for electric cars, in addition to exempting the vehicles from road tolls and allowing them to use bus lanes to avoid congestion.[37] Conversely, it taxes GHG-emitting vehicles very heavily. The combination of those policies led electric vehicles to account for 30 percent of Norwegian new car sales by the late 2010s. The US federal government, states such as California, and various utility companies have likewise offered large subsidies for the purchase of electric vehicles, often totaling more than $10,000 per vehicle in tax credits and rebates. Social equity questions around these programs still need to be addressed—they should not just represent subsidies for affluent individuals—but this can be done, for example through policies inversely scaling the size of the subsidy to household income.

Wealthy nations have the ability to offer massive subsidies for electric vehicles and other alternatives such as public transportation if they choose. For a similar cost as its 2017 tax cut the US could provide an average $12,000 subsidy for every one of the approximately sixteen million light-duty vehicles sold annually—incentivizing a rapid transition to an all-electric fleet. Another tool would be a feebate system, under which purchasers of GHG-emitting vehicles would pay surcharges while electric vehicle buyers would receive rebates. This policy wouldn't require public subsidy at all. With the shift to electric vehicles there are concerns in the US about how highway repair will be funded, since these funds currently come from gas taxes. However, new revenue mechanisms could be found, such as highway tolls or vehicle registration fees, which would have the benefit of disincentivizing driving and promoting more sustainable urban forms and healthier lifestyles.

A third way to speed the electric vehicle revolution is to make recharging easier. Many cities have already installed charging stations throughout their jurisdictions, typically in public parking lots and garages. Apps to help electric vehicle owners find easy recharging are also spreading. Tesla has its own charging infrastructure network to service its car owners. In just a few years, though, larger batteries and faster recharging technologies will make charging less of an issue. Drivers may be able to pull into recharging stations and fill their car batteries in twenty minutes or less while they shop or get coffee.

Limit Air Travel

Air travel is a rapidly rising source of GHG emissions (around 5 percent worldwide) that is not yet well incorporated into policy, equity, and sustainability frameworks. Passenger air miles rose at an average rate of more than 5 percent annually in the late twentieth and early twenty-first centuries before the Covid-19 pandemic, with a similar growth in air freight.[38] Travel by aircraft was often the single largest contributor to carbon footprints in affluent households.

Limiting air travel either through price or (more equitably) by allowing every individual a maximum number of trips per year may be necessary. In Sweden a movement called "Flygsham" or flight shaming arose in 2018 to put pressure on people to limit their air travel. Greta Thunberg highlighted the climate impacts of flying by refusing to travel this way and instead sailing across the Atlantic Ocean on a supporter's boat to attend the United Nations Climate Conference meeting in 2019.

Although more efficient jet engines and lighter aircraft have made air travel more efficient, further engine improvements are expected to yield only around 20 percent further energy savings.[39] Biofuels may potentially substitute for kerosene as jet fuel, but are more expensive and have environmental impacts of their own. No technical solution appears likely to completely eliminate the climate impacts of air travel. In the 2010s the aviation industry developed a Carbon Offset and Reduction Scheme for International Aviation (CORSIA), through which international flights would document their GHG emissions and purchase emissions offsets starting in 2027.[40] However, this agreement relies on offsets and appears a tokenistic effort by the industry to avoid more serious carbon taxes. More encouragingly, KLM, the Dutch airline, has developed a campaign called "Fly Responsibly" that encourages passengers to consider alternatives for short-distance flights.[41]

Limits to air travel could be achieved through strong carbon taxes, taxes or fees on takeoffs and landings, policies to restrict short-haul flights between cities when less carbon-intensive alternatives such as trains are available, tax changes so that businesses could not write off large amounts of air travel as a cost of doing business, or limits on the number of flights that a given individual could take each year. Most of these changes would need to take place at national or state levels. During the Covid-19 pandemic, air travel was greatly reduced as people everywhere learned how to accomplish many of same tasks as before virtually. Carbon emissions from air travel dropped by 60 percent.[42] The lessons could greatly help societies reduce the need for air travel.

A public discussion on how to limit and/or ration air travel seems imperative. Seeing distant family members and learning about other parts of the world might rise to the top as the most beneficial uses of jet airplanes. Regular commuting by air, whether to workplaces, conferences, or vacation homes, is far more questionable. As with so much else, such discussions return to questions of lifestyles and values. Minimizing air travel is only one of the sustainability benefits to be gained from living more locally and caring for the communities we inhabit.

Place a Rising Price on Carbon

To encourage a rapid shift toward climate neutrality throughout economies, an essential strategy will be to put a rapidly rising price on carbon. Some nations instituted carbon taxes beginning in the 1990s, although not nearly at the levels needed.[43] Finland was the first in 1990, with a tax equivalent to thirty-five Euros per ton of carbon dioxide-equivalent (CO2e), and other Scandinavian countries followed shortly after. Costa Rica began a tax of 3.5 percent of the market value of fossil fuels in 1997. Chile adopted a carbon tax equal to five dollars per ton, Denmark a tax equal to thirty-one dollars a ton, and Switzerland a tax equal to sixty-eight dollars a ton.[44] At the municipal level, the city of Boulder, Colorado, adopted a surcharge on electricity consumption in 2006 to fund the city's Climate Action Plan.[45] The tax raises about $1.8 million a year with a maximum residential rate of half a cent per kilowatt hour and has helped Boulder reduce its emissions 21 percent from 2005 levels.[46]

However, taxes below thirty dollars per ton of CO2e generally have only a small effect on fossil fuel consumption. To move rapidly toward climate neutrality these levies need to rise annually to much higher levels. Canada's twenty-dollar carbon tax, adopted in 2019, increases by ten dollars a year until it reaches fifty dollars in 2022, and is coupled with rebates to households. But still higher prices are needed. Economists argue that the price of carbon should rise until it reaches the marginal cost of damage done by each additional ton in the atmosphere.[47] But who can really calculate the cost of global warming? All we know is that the environmental, human, and economic costs are enormous and that it needs to be stopped as soon as possible. So this price signal needs to be strengthened and extended worldwide until climate neutrality is reached. Pricing also needs to account for equity concerns to ensure that low-income people and countries are not regressively taxed.

Cap-and-trade systems are a different way to use price to reduce GHG emissions. Governments set an overall cap on emissions that lowers over time and establish a limited number of permits to produce emissions within that cap. They then give away these permits or (preferably) sell them at auction. If companies need additional permits, they must buy them from one another, establishing a market in carbon credits. Australia, New Zealand, South Korea, China, and India have experimented with such emissions trading systems, although Australia's has been repealed. California has had a cap-and-trade system in place since 2012, although the price per ton of GHGs at auctions has usually been low, in the fifteen- to seventeen-dollar range. The European Union initiated a cap-and-trade program for industry in 2005. This system endured numerous problems, including excessive numbers of free permits given away and very low prices. The EU has made repeated fixes, most notably to reduce the cap limit by 2.2 percent per year.[48] Eventually this may establish a price signal that makes a difference in terms of a climate neutrality goal.

Whatever the mechanism, the impacts of higher energy prices on disadvantaged populations will need to be offset. This can be done in many ways. Funds collected through carbon taxes or cap-and-trade systems can be directly rebated to poor households through checks or tax credits. Or these revenues can be used to subsidize public transit and low-income home energy efficiency, thus saving poor households money. The Green New Deal proposed by progressive legislators in the US aims at simultaneously making broad changes in energy policy, tax systems, minimum wages, and social services so as to both move toward climate neutrality and benefit low-income populations.[49]

Adopt Climate-Friendly Lifestyles

Current GHG inventories for cities, states, and nations typically tabulate emissions by economic sector. They capture emissions from goods and services produced within a jurisdiction but not from those consumed by its residents. In wealthy countries more than 30 percent of emissions stem from people's personal consumption, including diet, and so are not counted within traditional emissions inventories.[50] These emissions are in effect exported to other places that produce the goods being consumed. Taking responsibility for these emissions and reducing them will need to be part of any climate neutrality program. This will mean rethinking consumption patterns.

Changing behavior and lifestyles, discussed further in chapter 9, is an as-of-yet underexplored dimension of climate action. However, many behavioral changes can both improve quality of life and reduce GHG emissions. For example, driving less and walking or biking more can improve health while lowering personal carbon footprints. Eating a more fruit- and vegetable-based diet has exactly the same effect. Greater sharing of material goods can promote community while reducing embodied emissions within one's possessions.

Move toward Climate-Friendly Agriculture and Food Systems

Although diet is a form of consumption, its emissions are so significant than they merit their own discussion. Agriculture accounts for at least 14 percent of global GHG emissions, and perhaps as much as 25 percent when land use changes and supply chains are included.[51] One study has estimated that the average British diet is responsible for an astounding 19.4 pounds (8.8 kilograms) of carbon-dioxide-equivalent emissions every day.[52] Diets in other developed nations are likely similar.

Luckily, there are many ways to reduce the dietary portion of our carbon footprint. By not eating meat, the average UK resident could reduce his or her food-based emissions by 35 percent. (Emissions associated with meat include methane from the decomposition of manure, methane from the belches and flatulence of cattle, and emissions from growing the food that livestock consume.) By eating chicken and pork rather than lamb and beef, people could reduce their food-based emissions by 18 percent. Vegetarian diets would yield even greater savings. By cutting down on food waste, for example by keeping more careful track of leftovers, people could reduce their emissions by 12 percent. And by eliminating hothouse-grown food or products flown to Britain from elsewhere they could lower their food-based emissions another 5 percent.[53]

Many initiatives promoted by governments at multiple levels can reduce GHGs from agriculture. These include better manure management, reduced use of nitrogen fertilizers, and steps to reduce and manage methane emissions from ruminant cattle. Reducing food waste is also crucial. Project Drawdown argues that a third of the world's food is never eaten and that this alone accounts for 8 percent of global GHG emissions.[54] Other food system changes such as reducing the control of food distribution by a few large multinational corporations and decentralizing agricultural land ownership worldwide will likely be needed as well. The public sector and NGOs can undertake social marketing campaigns to emphasize that eating mainly vegetables, fruits, grains, and nuts while avoiding processed foods is both healthiest for people and lowest-carbon. Changing diets in this way would save the US $114 billion annually in health costs, not to mention saving thousands of lives.[55]

Social equity questions enter into diet discussions as with most other topics. Many working-class and low-income people lack access to healthy food options. Cities can proactively identify the needs of specific communities within their borders and can seek to ensure that these groups have good access to markets supplying healthy and affordable foods, as well as incomes sufficient to buy it. Raising wages across the bottom half of the income distribution and ensuring that social safety net programs provide food security thus become important public policy goals.

Sequester Carbon in Ecosystems

Because all of the above initiatives together are unlikely to totally eliminate human GHG emissions, carbon sequestration will be necessary to reach climate neutrality or better. Large-scale reforestation at both tropical and temperate latitudes is one way to do this. Much-touted "Trillion Tree" programs are quite likely infeasible, ecologically counterproductive, and likely to distract attention from more urgent priorities.[56] However, smaller-scale, more context-sensitive reforestation programs can potentially improve ecosystem health in ways that remove carbon from the air and lock it within long-lived vegetation. New rangeland management practices that restore deep-rooted perennial grasses and limit grazing can potentially store carbon within soils. Although reduced agricultural tilling and better farmland management are often seen as ways to sequester large amounts of carbon, recent research shows that these are unlikely to be a panacea.[57] GHG emissions from agricultural soils are highly variable, and no single strategy seems likely to dramatically lower them.

The fossil fuel industry has for many years promoted the possibility of industrial carbon capture and sequestration (CCS), in part to rationalize the continued use of fossil fuels. Carbon dioxide from fossil fuel plant smokestacks would be captured, condensed, transported to appropriate locations, and then injected deep underground, where it would remain for many thousands of years. Other technologies might capture CO_2 directly from the air and sequester it underground.[58]

This industry vision is highly problematic. Most test projects begun in the 2000s were scaled back or canceled in the 2010s. It is uncertain whether carbon dioxide stored in geological formations will remain underground for the required amount of time. Capturing carbon from smokestacks and storing it takes energy, which reduces the overall efficiency of industrial processes. Approximately 30 percent of the fuel burned at a coal-fired power plant with CCS would be needed just to run the CCS system.[59] Partly as a result, the cost of such technologies would be high, especially when compared to the rapidly falling price of renewable energy.

If societies were willing to pay the cost, generating electricity by burning biomass with CCS could result in negative emissions from the electric power sector.[60] The plants and algae consumed in this way would take carbon out of the air that would then be stored for millennia underground. However, costs and technical difficulties are high, and avoiding emissions in the first place is a much more sustainable approach. The cheapest and most useful form of carbon sequestration is likely to be through biological systems, especially forests and rangeland, in which natural systems do the capturing and sequestration.[61]

Consider Geoengineering as a Last Resort

One last-ditch strategy that technological enthusiasts propose is geoengineering. Advocates of this approach back major interventions in the Earth's systems to reduce solar radiation and keep the climate tolerable while humanity lowers its GHG emissions and/or sequesters carbon.[62] Mechanisms include injecting sulfur into the stratosphere (blocking a percentage of sunlight for a number of years, as can happen when a large volcano erupts) or using iron filings to seed algal blooms in the ocean (thus soaking up carbon and depositing it as limestone when small marine creatures die and their bodies drift to the ocean floor). Or, seemingly out of the realm of science fiction, large pieces of very light-weight fabric could be placed between the Earth and the sun to reduce solar radiation.[63]

Such approaches would be enormously costly and would raise substantial risks of unintended side effects to the planet. Many ethical questions exist surrounding who has the right to artificially reengineer the Earth in this way or how nations would agree to do this. Geoengineering also raises the risk that societies would think that such technical fixes alone would be enough to avert catastrophe. So this approach should be kept off the table if at all possible.

. . .

Many more specific actions could be mentioned as well. One relatively easy step that local governments are taking is to cap landfills so that they don't release methane. Higher levels of government will need to move more quickly to require industry to replace existing refrigerants

(CFCs and HFCCs) with other versions that don't have powerful global warming effects. They will also need to require industry to emit less black carbon that can reduce the reflectivity of snow and ice in northern latitudes, causing the planet to absorb more solar energy.

Paul Hawken and his colleagues highlight a more extensive list of one hundred strategies for both lowering emissions and reducing atmospheric GHG concentrations in their Project Drawdown materials.[64] Some of their approaches are very specific ("In-Stream Hydro," "Bamboo," "Smart Glass") and some are more general sustainability goals that will have major eventual GHG reduction impacts ("Educating Girls," "Indigenous Peoples' Land Management," "Forest Protection"). One controversial recommendation is nuclear power. While acknowledging that nuclear fission is a very expensive way to boil water in order to run turbines and produce electricity, they argue for the continuation of existing nuclear plants and even a modest expansion of this industry as a bridge technology while the world develops other strategies. Others disagree, pointing out that nuclear power is highly uneconomical compared to energy conservation and efficiency programs or renewable energy, in addition to its many environmental risks.[65]

Although Project Drawdown does not mention it per se, reducing poverty and improving economic equality worldwide are among the most long-term important steps to reach climate neutrality and are the right thing to do for many other reasons as well. True, as residents of developing nations enter the middle class they are likely to increase their consumption and improve their diets. But if low-carbon lifestyles are more widely enabled, for example by making 100 percent renewable electricity and healthy, low-carbon foods universally available, higher standards of living need not result in rising GHG emissions. Poverty itself causes significant long-term emissions in that poor people cut down forests for firewood or farmland, have more children, drive old and polluting vehicles, and are often unable to access energy-efficient technologies.[66] In this area, as in many others, progress toward one global Sustainable Development Goal (SDG 1: No Poverty) can enable progress toward others (in this case SDG 13: Climate Action).

THE SOCIAL ECOLOGY OF CLIMATE ACTION

Many of the above climate neutrality strategies are daunting politically, culturally, and institutionally. However, technology and cost are usually *not* the main obstacles. Many emissions-savings technologies have existed for years and would save humanity money if widely adopted.[67] Rather, the problem is that our social and political institutions have been unable to bring about the necessary changes. People are also resistant to changing their lifestyles, even if relatively simple changes such as walking more or eating less meat would benefit themselves and the climate. To get around both obstacles, changes in social ecology are needed.

Social ecology changes will be discussed throughout this book. They include putting capitalism under firmer public sector control or phasing it out altogether in favor of alternatives such as a market-based non-profit economy (chapter 3). They also include greatly reducing inequality (chapters 4 and 5), reducing the influence of money over democracy and the media, and ensuring a much better educated and involved citizenry (chapter 11).

Even if capitalism persists in its present form, nations may need to nationalize fossil fuel companies to reduce their influence over political systems and convert them to other lines of work. Although prices fluctuate, estimates are that the US government could acquire a controlling share in the country's energy industry for $350–400 billion and that the British government could do likewise for $50–60 billion.[68] The public sector may also need to take over private electric utilities so that these companies do not prioritize the needs of investors and managers over climate neutrality and public safety.

Informational and educational strategies will be essential to change social ecology so as to support the move toward climate neutrality. Energy efficiency ratings of appliances are a start in this direction, but comprehensive labeling of embodied GHG emissions in products and services throughout the economy could help consumers become aware of the climate impacts of their purchases. Monitors showing real-time energy use within buildings could help change user behavior, much as dashboard displays help condition drivers of electric and hybrid vehicles to reduce energy use. Companies could be required to publicize net GHG emissions of their operations. Building entrances might feature colorful LED displays showing year-to-date energy used and generated.

Just as public health campaigns have helped reduce smoking in many countries, social marketing campaigns could promote climate-neutral lifestyles. School curricula, particularly at K-12 levels, could train the next generation in fossil fuel–free living. Knowledge of climate science and climate neutrality strategies could be made a basic requirement for degrees at all levels. In such ways, education—the process through which humanity trains itself to understand the world and act within it—could help produce a more constructive social ecology able, in turn, to address the climate crisis.

CONCLUSION

Getting to climate neutrality by the middle of the twenty-first century will be one of the biggest transitions humanity has ever gone through. Quite likely it will be larger and quicker than either the Industrial Revolution or the Agricultural Revolution. It will require profound changes in human behavior and social ecology as well as in ethics, public policy, economics, and technology. It will also require profound changes in cities—for example, more compact and balanced land use patterns to reduce driving; high-quality and equitable schools in cities and suburbs, streets that are friendly to bikes, pedestrians, and public transit; buildings that generate energy and are adapted to the local climate; and persistent social messaging about climate-appropriate lifestyles.

The transition to climate neutrality will likely result from some combination of awareness, education, leadership, public activism, an increased sense of urgency due to global warming impacts, growing recognition that structural changes in economic and political institutions and investments are necessary for many other reasons as well, and evolution of technologies such as photovoltaics and electric vehicles. Of course, it's possible that humanity will fail this growth test. But young people worldwide are stepping up. Let's each of us do what we can to bring this transition about, even if it means making substantial changes in our own lifestyle and becoming more politically engaged to demand institutional action.

2

How Do We Adapt to the Climate Crisis?

Even as communities move toward climate neutrality they will need to adapt to a radically changed global climate. The state of Florida will be a fascinating laboratory in this regard. Much of its coastline will flood regularly after one to two meters of sea-level rise this century.[1] Some six million of the state's people live in low-lying coastal cities such as Miami and Fort Lauderdale, which will be inundated by storm surges. A massive resettlement will be necessary, probably combined with new infrastructure to lessen the impacts of flooding.

Fort Lauderdale has already raised some roads to protect them and is concerned that its aquifers will become salty, threatening the city's water supply. The city admits that "incoming sea water adds salt at the edges of the Biscayne Aquifer and can infiltrate drinking water wells, puts pressure on pipes, increases the chance of breaks and requires additional maintenance. Infrastructure such as ports, schools, hospitals and landfills generally experience flooding at the borders and will require upgrades to on-site drainage systems. These issues are present with only a three- to seven-inch rise in sea level."[2] So it is clear that government officials are recognizing the enormous challenges that climate presents.

Sea-level rise threatens many other major metropolitan regions globally, including Alexandria, Shanghai, Rio de Janeiro, Osaka, London, and Tokyo. Countries like Bangladesh, Vietnam, and Thailand will have densely populated regions submerged. Worldwide, 640 million people live on coastal land below projected flooding levels by 2100.[3] As many as 13 million people in the US may need to relocate.[4]

Global warming impacts—among them more intense rainfall, stronger hurricanes, flooding, drought, sea-level rise, more widespread wildfires, crop failure, heat impacts, and changing disease vectors—represent a large and growing cost for societies. Hurricane Sandy alone caused $62 billion in damage to the New York region in 2012 and required $14.7 billion in federal aid for recovery and preparation for future storms.[5] The US General Accounting Office estimates that global warming had already cost the US federal government $350 billion by 2017.[6] A UN panel estimates that by 2040 the costs of the climate crisis could be $54 trillion.[7]

Global warming also represents the world's largest environmental justice problem. Hundreds of millions of poor and otherwise disadvantaged people will suffer because more affluent and politically powerful populations haven't reduced their GHG emissions. The people who have contributed the least to global warming are the most vulnerable, will bear the greatest costs, and are the least able to adapt. Climate justice impacts can already be seen within almost every climate-related event. The 780,000 New Orleans–area residents displaced by Hurricane Katrina in 2005 were disproportionately Black and low-income.[8] When Hurricane Maria hit Puerto Rico in 2017, it devasted communities containing some three million people of color, many of whom had to live without electricity, health care, or secure access to food for months. When torrential rains flooded Peru that same year, they left one million people homeless.[9] Social equity dimensions give even more urgency to climate adaptation.

Some climate adaptation steps will be relatively simple, like expanding urban forestry to increase shade and creating new parks in low-income neighborhoods. But others will need to be more radical, like retreating from coastlines or agreeing upon national and international plans to relocate millions of people. Even something as simple as urban tree planting has complex trade-offs, since shade-producing trees can consume scarce water, cause sidewalk upheaval and damage to property, create liabilities, shade roofs, and reduce the potential for on-site photovoltaic production. As with other issues, understanding the best sustainability strategies for any given situation requires contextual understanding, public involvement, and collaboration. It means recognizing the crisis and coming up with creative ways to adapt and respond. Yet climate adaptation is an opportunity as well as a crisis—a chance to imagine safer, greener, more equitable, and healthier communities for ourselves. How do we begin?

The most fundamental shift will be mental—away from the BAU attitude that the Earth is a stable, limitless system that humans can exploit, and toward a new understanding of humans as interdependent with the Earth's systems and each other. That may sound simple, but it's not. A main reason for the extraordinary strength of climate denial in countries like the United States, and general lack of climate action everywhere, is reluctance to give up the underlying belief that humans have the right to exploit and profit from the Earth. This anthropocentric attitude, like racism, sexism, nationalism, and other forms of parochial self-interest, lies at the root of many sustainability problems and needs to be tackled head on.

The climate crisis, as Naomi Klein argues, "changes everything."[10] Values and worldviews must evolve. Part of this change will be viewing climate adaptation as a challenge of positive social transformation. Adaptation cannot just seek to protect exploitative societies. Instead, it demands reimagination. It must create new, just, supportive, and equitable societies that live in balance with nature. We need, not just sustainability, but "just sustainability," as Julian Agyeman has argued.[11] To echo Albert Einstein, climate adaptation cannot take place with the same values and social structures that led to the problem in the first place.[12]

STRATEGIES FOR ADAPTATION

The paramount adaptation strategy, of course, is mitigation: quickly reducing GHG emissions to net zero and sequestering carbon from the atmosphere through ecosystem management practices. If the planet keeps warming indefinitely, all the adaptation programs in the world won't make cities livable. But while societies are reducing their emissions, many specific adaptation actions are also needed, integrated with mitigation and climate justice activities (table 2).

Anticipate Climate Justice Needs

Serious efforts for climate adaptation must start with social equity, since much of the harm that will result from global warming is due to ways that systematically disadvantaged communities worldwide have been left in highly vulnerable conditions. Addressing all forms of structural inequality within societies—not just climate vulnerability—should be

TABLE 2   STRATEGIES FOR ADAPTING TO THE CLIMATE CRISIS

Strategy	Description
Anticipate and address equity needs.	Reduce inequities generally in cities and societies. Proactively identify and address climate justice issues and vulnerabilities. Share adaptation resources globally and consider climate reparations.
Build social capital.	Strengthen ties between people so that we all can look out for, protect, and support one another in difficult times.
Cool communities.	Use vegetation, especially trees, for shade; light-colored materials to reflect solar radiation; multistory buildings, arcades, and courtyards to produce shade for human comfort.
Prevent flooding and storm damage.	Preserve floodplains and wetlands; use ecological strategies such as swales and permeable paving to hold stormwater on-site.
Minimize fire risk.	Avoid building in fire-prone areas; reduce fuel loads; use nonflammable materials for building exteriors; eliminate potential fire-starters such as aboveground power lines.
Use water wisely and prepare for drought.	Conserve water and improve efficiency of use; reuse and recycle all forms of water, including rainfall, gray water, and sewage.
Ensure public health.	Provide cooling centers and public education; ensure access to safe drinking water; research and respond to new disease threats.
Create more resilient food systems.	Develop local food system plans; increase local production; develop more varied distribution mechanisms such as community-supported agriculture (CSAs); improve water efficiency of agriculture; work toward more equitable land distribution, especially in developing countries.
Retreat from sea-level rise.	End development in flood-prone locations; remove development and restore coastal ecosystems; protect infrastructure; raise buildings; relocate populations; improve dikes and barriers where necessary.

the foundation for climate adaptation. If large sections of the population are poor or otherwise disempowered, they will have little ability to respond after disaster strikes and will suffer unnecessarily. In the case of New Orleans, addressing the situation of the African American lower Ninth Ward with its 34.7 percent poverty rate should long have been a no-brainer for any decision maker in Louisiana.[13] But this situation had existed for generations without action, leading to disproportionate suffering when Katrina struck.

Steps to identify specific risks to disadvantaged groups are crucial as well. Climate risk assessments are now widely conducted by local, state,

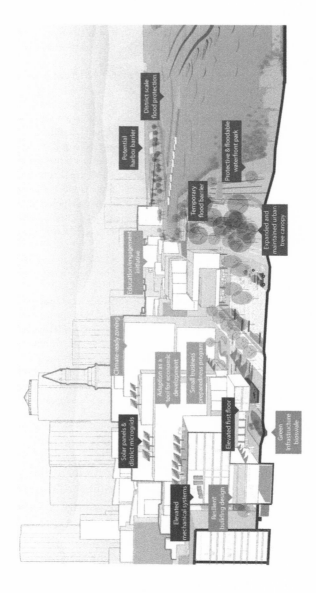

FIGURE 4. Boston's Climate Ready Boston plan combines a large number of strategies and identifies eight disadvantaged neighborhoods at special risk. (City of Boston, Climate Ready Boston Final Report, 2016, www.boston.gov/departments /environment/preparing-climate-change.)

regional, and national governments and can be combined with equity indexes to prioritize adaptation investment based on equity. Risk assessments typically summarize knowledge about how changing temperatures, precipitation, storm patterns, rising sea level, and disease vectors will affect a geographical area, then identify impacts on infrastructure, ecosystems, and at-risk populations, and finally highlight mitigating actions. For example, the City of Tacoma, Washington's Climate Change Resilience Study identifies its West End and North End as at higher risk from floods and landslides and recommends specific actions within these neighborhoods.[14] Boston's Climate Ready plan identifies eight neighborhoods within the city at special risk because of poverty, poor health, and limited English proficiency in combination with climate hazards such as flooding and heat.[15] This plan outlines actions ranging from flood barriers to small business preparedness (see figure 4).

However, once risks have been identified, action is essential. This will take leadership and equitable prioritization of funding for protective infrastructure. Studies beforehand accurately predicted the risks from Hurricane Katrina, Hurricane Harvey (which flooded Houston in 2017), and a Covid-19-like event.[16] Yet politicians and administrators chose not to act on these warnings, in part because of the political difficulty and cost. Stronger leadership that helps the public understand the need to pay for climate preparedness is urgent.

Build Social Capital

Improving the ability of people and local organizations to help one another is another essential climate adaptation step. Both before and after a disaster, these personal relationships and social capital are often more useful than slow-moving government programs and indeed can often point the way toward more cooperative social structures, as Rebecca Solnit argues in *A Paradise Built in Hell*.[17] *Social capital* refers to networks of trust, reciprocity, and cooperation between people[18]— ties that bind individuals together. If people in a community know and care about their neighbors, they will be more likely to look out for them and help them in time of need.

In countries that lack efficient governments and strong social safety nets, traditional practices have often created frameworks of mutual assistance. Within Islamic societies, for example, the concept of *Wapf* calls for wealthy individuals to leave large amounts of property to benefit less privileged people in their community, literally deeding the land, buildings, or money "to God" and charging a trustworthy individual to manage it for public benefit. This traditional endowment has helped many communities survive difficult times. Local governments can seek to reinforce such traditions.[19]

Building social capital is not easy. It depends on the overall health of the social ecology—things like the strength of local institutions; the extent to which people interact in daily life; the existence of shared activities and rituals; shared empathy and trust; and the strength of locally owned businesses. Steps to reduce inequality and segregation will help (see chapters 4 and 5). So will efforts to build local environmental stewardship and a locally oriented economy. Boston's "Climate Ready Story Project" reflects such a philosophy. One part of the city's larger adaptation plan, this project's website and events share stories of social resilience and connectedness emphasizing ties among neighbors, small businesses, and families.[20] Telling stories of collective challenges and local resilience allows us to humanize the impacts of climate and connect adaptation initiatives with our everyday lives.

Cool Communities

The effects of heat in any given place are complex and depend on variables such as humidity, elevation, vegetation, the quantity of paved surfaces, and air pollution.[21] They also depend on the underlying health conditions of the population, social equity factors such as whether poverty forces individuals to work long hours outside without protection from heat, and access to protective amenities such as air conditioning, pools, parks, and cooling centers. Whatever the context, heat can kill. One 2009 meta-analysis of thirty-six studies found a strong overall relationship between heat and mortality, with several reports suggesting a 3–6 percent increase in daily mortality in human communities for each 1°C (1.8°F) increase in temperature above thresholds of around 29°C (84°F).[22]

Heat need not be at extraordinary levels to be dangerous. Prolonged temperatures of 86°F (30°C) or lower, combined with high humidity, can stress the human body, especially for the elderly, young children, and those with medical conditions and comorbidities.[23] Such stress accumulates, particularly if there is no strong cooling at night, which is often the case for cities with high humidity. Cities in relatively cool, temperate climates can be most at risk from heat waves, since older dwellings in those places often lack air conditioning and good building insulation. Almost fifteen thousand people died from the August 2003 heat wave in France, where cities like Paris are full of older apartment buildings without air conditioning.[24] The Swedish National Health Agency estimates that seven hundred additional deaths occurred in that country during a heat wave in July 2018.[25]

Urban heat islands—the phenomenon through which cities with lots of hardscape absorb solar radiation during the day and become hotter than surrounding rural environments—receive a lot of press. Heat islands can make cities at least 3.6–9.0°F (2–6° C) hotter than nearby rural areas.[26] However, they very much depend on context, and in arid regions where surrounding deserts heat up and cool down rapidly in the course of each day, cities can actually experience an "urban cool island" with temperatures lower than surrounding territory. However, they are still hot. People can still sicken and die. What matters in the end is not whether the city is hotter or cooler than its surroundings, but heat itself and how well people are prepared for it.

Three main tool sets can help cool communities.[27] The first uses vegetation to provide both shade and evapotranspiration. Tree canopies provide both of these functions. Turf lawns provide only the second, while requiring more water and maintenance. Many poor neighborhoods and communities of color have far less of both than affluent districts where homeowners can afford extensive landscape design and irrigation. One study found that US neighborhoods that had been redlined in the past—a banking practice used to demarcate minority communities and deny home loans to buyers—had 37 percent less tree canopy.[28] Consequently many cities have undertaken tree-planting programs in recent years as a climate adaptation step. US cities such as New York, Chicago, Denver, and Los Angeles have all undertaken "Million Tree" urban forestry programs.[29]

However, the success of tree-planting programs at reducing heat is dependent on selecting the best species for the locations (and ones that will be able to adapt to changing climate), appropriate design (e.g., ample tree wells so that tree roots get enough water), maintenance, and in some places irrigation. Political, cultural, economic, and legal constraints often come into play. For example, in some cities homeowners are responsible for street trees and the sidewalk in front of their houses and may be less willing or able to take on tree maintenance and liability.

A number of communities have adopted regulations requiring private developers to plant trees to ensure that parking lots or other paved surfaces are shaded. Typically these regulations call for paved surfaces to be 50 percent shaded by trees within ten years as saplings grow to maturity. This approach makes urban greening part of the normal land development process, without the public sector having to launch its own program.

Vegetated roofs and walls can also help cool communities and have the further benefits of better insulating buildings and reducing stormwater runoff. After a 1995 heat wave was responsible for the death of more than seven hundred residents, Chicago became a leader in promoting green roofs. The city's zoning code offers a Floor-Area Ratio (FAR) bonus for developers who cover more than 50 percent of a building's roof area with plants. Those developers can then build higher or larger structures, gaining revenue. In its first few years this program resulted in 509 vegetated roofs, totaling more than 5.6 million square feet.[30]

The second tool set for cooling communities consists of high-albedo materials used for roofs, walls, and paved surfaces. These materials reflect solar radiation, keeping both buildings and the surrounding air cooler. Roof surfaces do not need to be white or even light-colored to do this. Since roofs account for about 60 percent of surface area in many urban regions, city or state policies to promote cool roofs can significantly reduce urban heating.[31] Cool roofs have been required by the California building code since 2005 and are promoted by cities such as San Antonio and New York. In recent years, phase change materials (which can temporarily store heat or coolness) have been shown to have cooling benefits apart from high-albedo materials by spreading thermal storage and release throughout the day or night.[32] On the ground, asphalt can be lightened by use of light-colored aggregate, dyes, or a light-colored concrete cover layer.[33]

The third tool set, least considered to date, is built form. Taller buildings spaced more closely together create more ground-level shade, which both cools the city and provides greater comfort for pedestrians. Awnings, arcades, and courtyards help create shade at ground level, allowing people to walk through hot cities without being in direct sun. They may also create stronger breezes within microclimates. Cities can require climate-adaptive built form through zoning codes, building codes, and design guidelines. For example, to maximize shade they can require minimum building heights and lot coverage, as well as reducing street width standards so as to have a greater percentage of the street right-of-way shaded.

Vernacular urban design in warm regions such as the Mediterranean has employed such tactics for millennia. In countries such as Italy, Greece, and Spain, traditional urban landscapes consist of continuous multistory buildings with courtyards along narrow streets. A downside, though, is that dense neighborhoods with narrow street canyons may cool down more slowly at night than landscapes in less dense cities, potentially increasing nighttime warming. This is usually less severe than daytime heat, however.

To combat the effects of heat on individuals, communities can set up cooling centers, usually in air-conditioned public buildings or religious facilities where residents can spend daytimes and/or nights to get respite from the heat. For example, Maricopa County, Arizona (which contains Phoenix), has set up fifty-three cooling centers providing services for more than 1,500 residents daily.[34] Cities can also increase the number of public swimming pools as well as vegetated parks and other types of cool public oases. Just creating such facilities is not enough, though. City heat response plans should emphasize warning systems, public education, programs focusing on high-risk populations, targeted outreach for the socially isolated, and transportation assistance.[35] Cities can also subsidize or provide free air conditioning for low-income residents as a part of climate justice initiatives. Alternatively, they can require landlords to install air conditioning in their rental units, as Montgomery County, Maryland, has done.[36]

Prevent Flooding and Storm Damage

Images of intense hurricanes often come to mind in relation to the climate crisis. Cities and towns can take multiple steps to reduce future damage from these. For example, one lesson learned from Hurricane Andrew in Florida in 1992 was to require builders to use screws instead of nails to hold roofs together. This greatly reduces the chance that the roof will fly off. Codes can also require builders to bolt buildings to their foundations, to use stronger doors and windows, and to hold beams together with strong hurricane ties.

However, although high winds may dominate the headlines, flooding is usually what leads to the greatest damage and loss of life within intense storms. This problem has worsened because of the growing intensity of rainstorms, hurricanes, tropical storms, and monsoons, as well as sea level rise and conversion of winter precipitation from snow to rain in some parts of the world. A warmer atmosphere can hold more moisture and energy, leading to situations in which unprecedented amounts of rain fall in a single day. Hurricane Harvey, which stalled over Houston in 2017, dumped more than fifty inches of rain on some parts of that region over four days—more precipitation than most of the world's cities receive in a year. With wetlands largely replaced by paved surfaces that rapidly shed water, the Houston area saw massive flooding, resulting in $125 billion in damages.[37] As a result the traditionally Republican city announced a climate action plan on Earth Day 2020 and acknowledged that "after three 500-year floods in as many years culminating with the largest rain event in North American history, climate change is an unprecedented challenge for Houston. Sustainability and resiliency go hand-in-hand and this plan is essential to the health and economic vitality of Houston's future. The time for bold action is now."[38]

Preserving or restoring floodplains through investments in green stormwater infrastructure (GSI) is essential to prevent flooding, since these low-lying areas of riparian vegetation absorb and slow stormwater. Over the past century floodplain development has occurred all too frequently as landowners and local governments have been reluctant to accept that future sustainability benefits should take precedence over their ability to make short-term profits and raise the tax base. City officials have been reluctant to declare parts of their communities "off limits" to development. In the US, government-funded flood insurance and disaster relief services have also helped rationalize development in flood-prone locations by removing the inherent financial risk. Ending insurance of new or existing development in such areas (with assistance for low-income residents to relocate) and taking steps to ensure that local governments are financially secure from other sources besides local property taxes would help discourage floodplain development. But this will require a reimagining of local government revenue systems and creative climate-ready policies that are centered on equity. New York State's Climate Risk and Resiliency Act of 2014 and Climate Leadership and Community Protection Act of 2019 may serve as models of state leadership to rethink climate risks and promote equitable investments in climate resiliency.[39]

Two opposing philosophies can be applied to specific flood protection mechanisms.[40] The industrial approach under BAU emphasized large, highly engineered interventions to control water flows—dams, levees, channels, and at times barriers across rivers or bays. The Dutch have used such means historically to reclaim land from the sea, and the US Army Corps of Engineers used them less successfully to protect cities, reduce flooding risk from rivers, and open up floodplains for development. This approach will have some limited use in the future, for example to protect existing neighborhoods by strengthening dams and levees, particularly if those areas are home to vulnerable populations with little ability to move. However, as a long-term strategy this approach is expensive and in many cases futile.

In contrast, the ecological approach to flood management seeks to work with natural systems rather than to control them. It avoids development in locations certain to flood and restores previously urbanized wetlands so they can slow and hold water. Associated with terms such as *GSI* and *Low-Impact Development (LID)*, this approach emphasizes holding and infiltrating stormwater on-site through use of permeable paving, swales, vegetation, green roofs, and fewer hard surfaces. The hydrologic system of an urban area can thus accommodate far larger amounts of stormwater. This philosophy underlies New York City's $19.5 billion climate resiliency plan adopted after Hurricane Sandy. It also underpins China's "sponge cities" program, begun after a series of floods in 1998, which uses green spaces and permeable paving in urban areas to absorb stormwater.[41] In addition to their function of preventing flooding, such systems have ecological, aesthetic, health, economic, and recreational benefits for communities and dovetail with the biophilic cities strategies discussed in Chapter 8.

Cities, states, and national governments can require all new development to retain or slow stormwater on-site in these ways. Oregon, for example, requires developments to keep 80 percent of average annual runoff volume on-site.[42] Pioneering projects such as the Village Homes neighborhood in Davis, California, have kept all stormwater on-site since the 1970s. In addition to such mandates, cities can incentivize ecological landscape design to reduce flooding. Washington, D.C., gives rebates of up to $6,000 to homeowners who replace pavement with pervious surfaces.[43] Seattle's RainWise program offers rebates to property owners for installing rain gardens (landscaped areas holding rainwater) or cisterns.[44] Auckland, New Zealand, promotes use of rainwater tanks to reduce stormwater flows and retain water for either nonpotable or potable (drinking water) uses.[45]

Communities with combined sewage and stormwater systems are at particular risk from flooding, since excess stormwater often causes sewage treatment systems to overflow and pollute waterways. Reducing floodwater volumes is even more important in these locales. Philadelphia's Green City, Clean Waters plan adopted in 2011 is a good example of municipal action in this regard. That city incentivizes private developers and citizens to add permeable pavement and create "living landscapes" to slow, filter, and infiltrate stormwater on-site. In the program's first five years developers added more than 1,100 landscape elements to handle floodwater in this way.[46] The challenge with this sort of initiative is that it relies on a distributed system of urban environmental management requiring landowners to partner with public and private entities. Important questions about capacity, ownership, funding, and long-term maintenance of projects need to be addressed.[47]

Increasingly intense rainstorms led Copenhagen to implement a climate resiliency program in the 2010s. One 2011 storm dropped six inches of rain in three hours and caused six billion kroner ($1 billion) in damage. The city focused on replacing hard surfaces in flood-prone areas with grass, swales, and tree planters; requiring that new flat-roofed buildings have green roofs to slow runoff; waterproofing cellars; creating additional retention ponds; increasing the size of sewers and pumping stations; and directing water away from buildings toward parks and other open spaces.[48] A climate-resilient neighborhood program added green spaces and used bicycle paths as stormwater channels to carry water toward the harbor, thereby simultaneously increasing resilience, recreation, and mobility.

Managed retreat from floodplains is a radical but increasingly necessary step that is gaining adherents. Homes in risky and flood-prone areas will almost certainly flood eventually. Why not deconstruct them now and move residents to higher ground, rather than waiting for disaster? We can proactively plan for the relocation of at-risk communities in thoughtful and equitable ways rather than doing so after tragedy strikes. Climate relocation has been employed in a number of locations worldwide, especially after coastal tsunamis.[49] For example, India relocated at least twenty-two thousand households along the coast of Tamil Nadu after the 2004 Indian Ocean Tsunami.[50] Eventually it may be necessary to relocate entire cities. This will be traumatic for many people and must be done in an equitable way, but it is the logical cost of past BAU practices.

Minimize Fire Risk

In addition to water, fire is an enormous climate danger, both by directly threatening people's homes and by creating smoke, which is a health hazard. Some three billion of the world's people live in dry locations—arid, semiarid, steppe, or Mediterranean climates. Fire is a natural part of many of these landscapes. Hotter temperatures, stronger winds, and more frequent and/or powerful thunderstorms due to global warming will increase fire risk. Strong winds, for example, quickly fan any small spark and can turn what would have been a small-scale, local grassfire into a raging firestorm. California, Australia, and Greece have all seen large-scale fires in recent years. Thousands of people were forced to evacuate their homes, and many lost their lives and livelihoods. The smoke from these fires is likely to lead to long-term lung damage.[51]

How can communities reduce the threat of wildfires? A number of relatively small steps can help. Keeping fuel loads reduced next to urban areas through selective pruning of trees and thinning of undergrowth is one of these. Reducing "fire ladders," through which grassfires can climb up into treetops, is another. Clearing brush away from buildings and requiring that buildings in fire-prone areas be clad with fire-safe materials rather than, say, wooden shingles is a third.[52]

Jurisdictions can also focus on reducing the human causes of fires. One basic need is to separate human dwellings from flammable woodlands so as to reduce the risk that sparks from electrical short-circuits or machinery will start fires. The 2015 Valley Fire in California, for example, was started by faulty wiring on a hot tub next to a vacation home in the middle of woodlands. It then burned through 119 square miles (308 square kilometers), destroyed the community of Middletown, killed four people, and caused nearly $1.5 billion in damage.[53] The state was forced to spend $57 million fighting the fire. With the ongoing worldwide phenomenon of rural sprawl—homes widely spaced in natural areas—the risk of such disasters is increasing. Local governments can reduce such risk by adopting strong subdivision and zoning controls preventing or restricting dispersed development in the wildland-urban interface. They can also require fire-smart development practices such as cleared buffers at the edge of neighborhoods.

Aboveground electric power lines in wooded areas are a related fire problem. A heavy windstorm can snap tree branches that fall on such power lines. Unless power is quickly shut off, the downed lines can generate sparks that ignite dry grass and shrubs. Governments can require that utilities clear trees along the power lines to reduce risks, but not putting aboveground lines and the developments they serve in such locations in the first place is an even better strategy.

Use Water Wisely and Prepare for Drought

Water availability and quality are a growing challenge in many parts of the world because of population growth, excessive human use, unsustainable farming practices, and pollution and will become a far greater problem with global warming. Large cities such as Cape Town, São Paulo, Bangalore, Cairo, Jakarta, and Mexico City have already run dangerously low on safe drinking water.[54] By 2071 nearly half of the US's 204 freshwater basins will not be able to meet the monthly demand for water.[55] Groundwater aquifers in places like the western US are often left over from the last ice age when climates were far wetter and are being depleted by human use. Glaciers and snowcaps that currently feed rivers ranging from the Ganges to the Colorado are melting. Climate zones are likely to migrate north, bringing Sahara dryness into southern Europe and further parching the American Southwest, leading to food and water shortages as well as large-scale migration, some of which we are already witnessing.[56]

What can be done to ensure that communities have adequate supplies of water? First, as with any resource, we need to conserve the resource and improve the efficiency with which it is used. In particular, agriculture, the biggest water user in many parts of the world, can be made much more efficient. Drip irrigation or microsprayers can be substituted for flood irrigation, and monitoring soil moisture and watering only at night can help conserve this liquid resource. Conservation tillage, mulches, and cover crops decrease evaporative water loss. Farmers can build ponds to store rainwater for use throughout the year. Crops can be grown that require less water, for example substituting sunflowers, safflower, millet, or olives for alfalfa and rice.[57]

There is no good reason to put this precious resource literally down the drain in our homes. Many jurisdictions worldwide already require low-flow showerheads, faucets, and toilets in new development. These should be mandated everywhere. Although they historically used at least 3.6 gallons per flush, many toilets now on the market use less than one gallon. Gray water within buildings can be reused to flush toilets, while rainwater harvested from roofs and pavement can be used for irrigation. Cities can require that existing water fixtures be retrofitted to current standards at time of building sale.

In dry locations sewage can be recycled into drinking water after being put through a three-stage treatment and filtration process. After all, every molecule of water we drink has previously been through the bodies of countless past organisms, and much municipal water comes from rivers in which many creatures have spent their lives. Windhoek, the three-hundred-thousand-person capital of Namibia, is one of the world's pioneers in water reuse technology. Since 1968 it has recycled sewage into drinking water through a multistage process that includes purification with ozone and activated charcoal. Visitors from other countries flock there to learn more about the process.[58]

Increasing the supply of water is usually not necessary if conservation, efficiency, and recycling steps are undertaken and is rarely desirable for environmental and cost reasons. New dams on rivers negatively affect fish, entire ecosystems, and human communities living nearby. New wells may deplete aquifers, cause neighboring wells to go dry, and lead to ground subsidence. Desalination plants are very costly and use enormous amounts of energy. Almost always efficiency and conservation are better options than expanding supply. With global warming, there will be tough decisions to be made about whether some communities still have the water supplies necessary to maintain their population.

Ensure Public Health

Besides heat, other types of public health threats will arise from global warming, each requiring its own response. Drought can also lead to problems of scarce or contaminated water. Floods can lead to localized problems with mold within flood-damaged structures or diseases such as cholera that result when the bodies of animals or humans languish in floodwater and associated pathogens contaminate drinking water. Fires can lead to breathing problems for downwind populations exposed to smoke. Governments can proactively identify and address such health challenges. For example, in fire-prone regions local or regional agencies can develop good air quality monitoring networks, real-time websites informing residents of smoke risks and ways to protect themselves, guidelines for shutting schools and businesses, and stockpiles of masks that can be distributed to vulnerable populations.

As climates change, pests carrying diseases will shift their ranges. Lyme disease and the West Nile virus are examples of diseases whose ranges may already have changed in North America because of global warming. Malaria and other tropical illnesses are likely to move to new, previously more temperate locales. Sweden has documented an increase in cases of vibriosis, a potentially lethal set of diseases caused by species of the bacterial genus *Vibrio,* which occurs in warm seawater and can cause gastroenteritis and wound infections.[59] Most notably, the global 2020 Covid-19 pandemic can trace its roots to environmental degradation. With disease geographies changing, public health professionals will need to identify disease vectors, risks, and control strategies. Vulnerable populations need to be identified and public education conducted.

Create More Resilient Food Systems

One of the biggest global climate risks is that food systems in part or all of the world will crash, leaving populations without sustenance. Those food systems are already failing to provide hundreds of millions of people with healthy diets daily. With heat, droughts, floods, and migration of agricultural pests, the number of communities affected may swell. Scientific consensus as reported by sources such as the Intergovernmental Panel on Climate Change (IPCC) is that in the short term warmer temperatures and higher CO_2 levels may be increasing yields of some crops in temperate regions but are decreasing average yields in tropical areas.[60] In the long term decreased yields are likely in both climatic zones. Disruptions in food systems are likely as yields, pest distributions, and climate change. Wars, mass migrations, and other social conflicts may result, as well as mass starvation.[61]

Community food systems plans that emphasize local food production and distribution for local consumption can help build food resiliency in the event that far-flung global food supply chains are disrupted. Such plans often focus on developing local farmers' markets, better distribution networks for local producers, consumer-supported agriculture networks linking households to local farmers, and programs to support farming families and workers. For example, Asheville, North Carolina, has adopted a Food Policy Action Plan that emphasizes increasing local production, improving community nutrition, building networks of regional partners, and preparing for food emergencies.[62] Resilient food production can also be integrated into the design of new communities. In one plan for a new district of Manila, the international firm Sasaki Associates considered such factors as preservation of farmland, maintaining pollinator function, and ways to integrate Filipino culture.[63]

Retreat from Sea-Level Rise

Perhaps the most radical forms of climate adaptation will relate to sea-level rise. For decades the IPCC and other authorities have been telling us that oceans are likely to rise on average one to three feet this century. However, these are conservative figures, based primarily on past rates of sea-level change during the last hundred years. Oceans have already risen eight inches with the relatively modest warming that has already taken place (~1°C or 1.8°F), and this rate will accelerate as melting takes hold of the large Greenland and Antarctic ice sheets. Recent modeling and paleoclimate data show that sea-level rise of several meters a century is possible—an inch a year or more.[64] Human preparations for sea-level rise will need to become more rapid.

Rising sea levels will make life increasingly difficult for coastal development. Main risks come not so much from the steady rise of mean sea level as from intense storm surges combined with high tides. As we have seen, these risks also include the intrusion of seawater into coastal aquifers and damage to coastal infrastructure including sewage systems, drinking water wells, roads, subways, power plants, and electrical lines. Hurricane Sandy's effects on New York City show how storm surge from a single event can overwhelm a metropolitan area. That single storm shut down the lower portion of Manhattan—one of the world's most intense centers of activity—for several weeks.[65]

Many of the same strategies that prevent flooding anywhere help minimize damage from sea-level rise. The starting point again is to prevent development in flood-prone areas, in this case coastal wetlands, barrier islands, and any coastal locations within a few meters of sea level. Restoring vegetation and natural systems in these areas can help protect low-lying coastal communities by absorbing floodwater from storm surges and buffering developed areas from wave action. Elevating buildings and moving electrical systems and other sensitive utility infrastructure out of the basement or bottom floor can improve building resiliency. New York learned this the hard way in Sandy, when ground floors and basements of many buildings in Lower Manhattan were flooded. Protecting subway entrances, sewage treatment plants, and low-lying roads from flooding can preserve functionality of this infrastructure. Seawalls, dykes, and levees may be necessary in places.

New York City's post-Sandy climate-preparedness blueprint, called *A Stronger, More Resilient New York,* is a leading example of a plan focused on sea-based threats.[66] The document details some 250 potential actions along the above lines, attaching price tags to each that total $19.5 billion. Although that sounds like a lot, it is far less than the economic damage caused by the storm. Luckily industrial countries have the money for climate adaptation. Completely funding the New York area's climate resilience plan, for example, would cost only about 3 percent of the $721 billion annual US military budget (in 2019), which exceeds those of the next ten largest national military budgets in the world combined.[67] Less developed countries are not so lucky in terms of resources, and international mechanisms will be necessary to assist them. More-developed countries will need to open their borders to some climate refugees and acknowledge their contribution to care for the global community.

An additional set of sea-level rise strategies can be grouped under the heading of retreat. Essentially, humans will need to move communities out of the path of sea-level rise. Governments can buy the homes of residents in low-lying districts, enabling them to purchase new homes in safer places. Or they can condemn those dwellings and directly provide the residents with alternative places to live. Or the residents themselves can individually or collectively make a choice to leave. If insurance companies decide not to insure properties in at-risk areas, that alone may be enough to bring development there to a grinding halt and to force existing residents to leave. Without insurance, property values will plummet and there will be little point to putting money into improvements or even maintenance. It will then become much easier for the public sector to acquire properties and return them to nature. All of these policies have serious equity implications that need to be planned for and addressed.

A basic question exists over who should pay for retreat, and how much. If private insurers will no longer cover at-risk properties, should governments do this themselves, as the US government has historically done for coastal properties? That may incentivize further development in risky locations, and bails out developers and property owners who took advantage of dysfunctional progrowth local politics to develop wetlands and coastal areas that should never have been built upon. Enormous amounts of money may be involved in retreat, and it will be important to make sure it is spent equitably and wisely. The US Department of Housing and Urban Development has already started a $16 billion program to relocate owners of flood-prone properties, and total US federal disaster recovery between 2005 and 2020 cost close to half a trillion dollars.[68] Humanitarian relocation assistance seems desirable for low-income communities but not top-dollar beachfront luxury homes. Public understandings of property rights may need to change. Landowners will need to realize that they do not necessarily have the right to buy or develop at-risk property if doing so harms ecosystems and imposes large potential costs on society. The public sector will also need to figure out ways that local government can be securely funded besides relying on property taxes and fees from inappropriately located, climate-risky development.

Local governments can potentially use zoning codes to encourage retreat from sea-level rise. Through a phased process that gives landowners many years of advance notice, they can rezone low-lying zones as nondevelopable coastal buffer while at the same time increasing housing densities for better-located, higher-elevation sites. Municipalities might establish sunset provisions for occupation of existing coastal properties, giving owners several decades to plan their relocation inland.

One interesting proposal in this regard (although not linked to climate change) is provided by Richard Register's late twentieth-century "ecocity zoning" proposal, in which large areas of existing urban areas would be zoned for ecological restoration—with existing structures gradually removed—while other portions of those communities are zoned for intensification.[69] Owners of land to be restored would be compensated by awarding them extra development rights in the intensifying neighborhoods, which could in turn be sold to developers. Financial and procedural mechanisms for such transfers of development rights (TDRs) have already been used by jurisdictions such as Montgomery County, Maryland, as a way to protect farmland from development.[70]

In some cases entire cities or nations may become untenable. As suggested at the beginning of this chapter, much of South Florida may be uninhabitable by the end of the century. Tragically, low-lying island nations such as the Seychelles are almost certain to disappear beneath the waves. Much of low-lying Bangladesh may become too frequently devastated by storms and tidal surge to be inhabitable. States, nations, and international agencies will need to step in to help populations in these places relocate and to assist other regions in accommodating climate refugees. Those nations most responsible for the climate crisis historically should be most responsible for taking on and funding this role as well as providing refuge. These actions might be considered climate reparations—payments based on acknowledgment of past responsibility.

CONCLUSION

Nothing is going to completely protect cities and towns from the impacts of a future global climate that will be many degrees hotter and will feature more extreme and unpredictable weather events than in the past. Still, careful attention to adaptation will lessen impacts and lead to many co-benefits (including in some cases GHG reductions).

Climate-adapted communities of the future will be far greener, with networks of vegetated public spaces and swales to absorb stormwater, as well as large, drought-tolerant trees (except in very arid regions) shading many streets, public spaces, and buildings. Buildings will be designed to stay cooler and produce more shade. Pavement will be lighter colored and pervious so as not to absorb heat and to allow stormwater to percolate into soils. Many roofs will be vegetated for cooling, building insulation, and water retention purposes. Buildings will be stronger. Some will be elevated to protect against flooding. Buildings' utility systems will be placed in attics or on roofs instead of in easily flooded basements.

Urban landscapes will look different. Large buffer zones will protect communities from waterways that may flood or woodlands that may burn. "Living shorelines" with dunes and wetlands will replace beachfront development. Buildings will be grouped together more densely in ways designed to maximize ground-level shade and minimize broad parking lots and street surfaces that would absorb and retain heat and cause stormwater runoff. Where appropriate, new levees, channels, and dykes will protect vulnerable communities while long-term relocation efforts proceed.

To reduce environmental justice impacts of climate change, leaders will proactively address social and racial equity problems and the needs of disadvantaged communities. Cities will identify vulnerable populations and plan to protect them. They will use big data and Geographic Information Systems (GIS) modeling to make planning predictions and assess cumulative environmental justice vulnerabilities. Public health personnel will advise planners and help prepare for many contingencies. They will ensure that vulnerable residents have cool homes and easy access to cooling centers during heat waves, and they will work with employers to protect farmworkers, construction workers, and others who must work outside (for example, by scheduling some work at night). They will also track the spread of disease-carrying organisms as climate zones shift and will proactively develop strategies to protect populations.

Globally, nations will mobilize massive amounts of international assistance to help poor countries most affected by the climate crisis. Populations will be relocated away from sea-level rise, sometimes to other countries that willingly accept these climate refugees. In these and other ways, humanity will be doing its best to adapt to life on a very different Earth.

3

How Might We Create More Sustainable Economies?

One of the economic success stories of the early twenty-first century—as conventionally construed—is Amazon. Within twenty years of its founding in 1994 it had become the largest retailer in the US. However, Seattle, Amazon's home, has seen extraordinary increases in housing prices and homelessness. Journalist Paul Roberts describes the city's dilemma: "Most would acknowledge the extraordinary prosperity that Amazon has brought to Seattle since Jeff Bezos and his start-up arrived in 1994. But they are also keenly aware of the costs, not least the nation's fastest-rising housing prices, appalling traffic, and a painful erosion of urban identity. What was once a quirkily mellow, solidly middle-class city now feels like a stressed-out, two-tier town with a thin layer of wealthy young techies atop a base of anxious wage workers."[1]

The lack of affordable housing became such a crisis that in 2016 Seattle residents agreed to tax themselves $290 million to build more of it.[2] In 2018, the city council decided to levy a tax on employers with more than thirty employees to raise $48 million yearly for programs to address homelessness. However, large companies like Amazon and Starbucks fought back and forced the city council to reverse its decision.[3] These large companies had amassed such political clout that the local government could no longer regulate them.

Cities are embedded in larger economic systems that determine their overall sustainability prospects and constrain their day-to-day policy choices. Reimagining sustainable cities means questioning the economic structures that have led to BAU. Should the future be a world run by companies like Amazon, ExxonMobil, and Starbucks and guided by the stock market and international financial speculation? How can concerns for social, racial, environmental, and economic justice guide economies instead?

Global capitalism has brought affluence to some but has helped create global warming, extreme economic inequality, systemic racism, environmental injustice, overconsumption, pollution, and dysfunctional democracy. Its emphasis on individualism and ever-growing consumption has dominated politics and constrained environmental policies. In an incisive but little-noticed volume nearly thirty years ago Martin O'Connor and others asked the question, "Is capitalism sustainable?"[4] Many contributors to the book concluded that it was not.

More and more of us believe in the need for economic system change. The Covid-19 pandemic and the Black Lives Matter movement have highlighted the inability of US capitalism to ensure public health or address structural racism. Even before these events a Harvard poll found that only 42 percent of Americans aged eighteen to twenty-nine supported capitalism, with 33 percent favoring socialism instead.[5] A 2018 Gallup poll found that Americans who identified with the Democratic Party had a more positive view of socialism than capitalism.[6] Such findings would have been unthinkable a generation ago. This questioning of economic and political systems is not unique to the United States. Populism in many nations is driven by working-class people protesting the fact that current economic and political systems are leaving them behind. Unfortunately right-wing populist politicians usually just reinforce BAU economic structures, making inequality worse.

Writers such as Harvard cognitive scientist Steven Pinker and Microsoft founder Bill Gates defend capitalism by arguing that the human condition is far better now than it has ever been, with less violence, more individual opportunity, and higher quality of life.[7] Capitalism and the Enlightenment emphasis on rational thought and science, Pinker believes, have brought us these benefits.[8] There is some truth to these arguments. Other pundits have argued that there is no real alternative to capitalism except the totalitarian versions of socialism represented by the former Soviet Union and its allies. Francis Fukuyama famously stated his belief in *The End of History* (1992) that liberal democracy and free-market capitalism represent the end point of human cultural evolution.[9]

Yet even Fukuyama has been forced to rethink his argument.[10] Concepts like "capitalism" and "socialism" have evolved for centuries and will continue to do so. There are no pure forms, and in a reimagined society we can mix and match them to meet environmental, social, economic, and equity goals. Societies worldwide illustrate various types of mixed economies, suggesting that improved versions of political and economic systems may be possible. Countries such as Sweden and Denmark are forms of social democracy in which capitalist markets thrive despite the presence of a large public sector and a strong social safety net. The US, conversely, far from representing a "free market," can be seen as socialism for the rich, in that many levels of government build infrastructure, supply services, provide tax breaks, tolerate monopolies, and otherwise frame the conditions within which private capital is accumulated.

Radically new forms of economy may also be possible. Why not a market-based economy populated by nonprofit rather than for-profit businesses? Why not an economy based on cooperatives through which workers, consumers, or producers own companies collectively? Why not economies in which capitalist businesses are limited in size and extent or are licensed under charters requiring transparency, public service, and worker and community representation on their boards? Why not a strict separation between corporations and governments so that the former do not corrupt the latter? In the early twenty-first century we are at the beginning of a process to reimagine and develop new economic systems that better support democracy, human welfare, and sustainable cities.

THREE POLITICAL ECONOMY PARADIGMS FOR SUSTAINABILITY

Any consideration of economics must start with big-picture political economy (the interwoven economic and political systems that structure societies). Three political economy paradigms are potentially useful in creating a context for sustainable cities: reformed capitalism, social democracy, and democratic socialism (figure 5). Other possible philosophies have been proposed historically, including anarcho-syndicalism (a decentralized, egalitarian, democratic system in which workers have primary control over economic production). But here we'll consider that a decentralized version of democratic socialism.

Reformed Capitalism

Those of us who live in the United States, Canada, Australia, and (in recent decades) Britain have experienced relatively laissez-faire forms of

Reformed capitalism	Social democracy	Democratic socialism
• Mainly private businesses	• Privately + publicly owned businesses	• Mainly publicly owned businesses
• Mainly for-profit organizations; some nonprofit or cooperative	• Mix of for-profit, nonprofit, and cooperative organizations	• Nonprofit or cooperative organizations; most controlled by the state
• Public sector strongly regulates business for environmental protection	• Public sector strongly regulates business + oversees public companies	• Public sector oversees public companies for environmental protection
• Strengthened social safety net	• Strong social safety net	• Strong social safety net
• Strong separation between monied interests and government	• Strong separation between monied interests and government	• Strong separation between monied interests and government
• Greatly reduced social inequality	• Reduced social inequality	• Low social inequality
• Civil society helps public and private sectors focus on sustainability and provides many essential services	• Civil society helps public and private sectors focus on sustainability; public sector provides most essential services	• Civil society helps public sector focus on sustainability; public sector provides almost all services
• A challenge of moving towards less materialistic and individualistic values and nurturing values of cooperation and caring	• A challenge of nurturing values of cooperation and caring	• A challenge of nurturing values of cooperation, caring, and individual initiative

FIGURE 5. Sustainable cities will almost certainly need a political economy paradigm that provides an alternative to recent neoliberal capitalism. Here are three possibilities.

capitalism in which neoliberal politicians have sought to reduce the public sector role and allow the "free market" to function. However, it is clear to most by now that markets are never really "free." The public sector plays an enormous role in establishing the conditions within which business operates. It educates workers, sets monetary policy, provides infrastructure, regulates markets and professions, and ensures public health, safety, and welfare. Government often intervenes to subsidize favored industries, leases the airwaves and other public resources to private companies, and at times bails out banks and industries deemed "too big to fail."

Moreover, a free market wouldn't necessarily be desirable even if it could exist. Capitalism doesn't do well at internalizing the externalities of production, including pollution, GHGs, and the exploitation of workers and communities. It also tends toward monopoly situations in which a few companies dominate key industries. Other problems include its tendency to concentrate wealth in a few hands and to undermine democracy as firms seek to use the government for market advantage. Also, capitalism does not do well at addressing past injustices or working toward equity. Hence a strong public sector role is always needed to control market activity and to ensure that collective values are advanced by private sector activities.

The question in terms of sustainability is not whether a stronger public sector role in regulating capitalism is needed (that is a given) but which forms it should take.[11] Various categories of reforms are necessary: those to encourage a carbon-neutral economy, those to improve social, racial, and gender equity, those to ensure affordable housing and health care for all, those to reduce pollution and bring about sustainable resource use, those to address inequalities between the Global South and North, and those to better insulate democracy and the media from control by capitalists. We discuss more specific strategies within these categories elsewhere within this chapter and volume.

Social Democracy

The social democratic paradigm, most prevalent in northern Europe, is capitalist in that most businesses are owned by private investors and operate within relatively open markets. However, the public sector regulates capitalism closely, provides a strong social safety net, and removes some human needs at least partially from the capitalist market (for example, health care, childcare, education, and housing).[12] The latter steps help meet social welfare and equity dimensions of sustainability. Social democratic countries typically also promote unionization of workers, creating a power base to counter the influence of corporate owners and managers. In the long run this arguably improves the income and working conditions of much of the population.

In line with their emphasis on social welfare goals, social democracies often try to create a collaborative political culture. Stakeholders (primarily business, labor, government, and organizations of civil society) work together to try to ensure that social and environmental goals are met as well as economic ones. In the Netherlands, for example, provincial and local governments have worked with companies, public-private partnerships (such as the large housing cooperatives that provide most rental housing), research institutes, and citizens' organizations to develop environmental and economic policy.[13]

Norway, Sweden, Denmark, and Finland are the prototypical social democracies, although Germany, the Netherlands, Belgium, France, and Britain (in the decades after the Second World War) might be included as well. Social democracies arose in these nations in response to the need to heal from the devastation of World War II and to create healthy societies that would not engage in future wars.[14] Although under increasing pressure from right-wing parties during the neoliberal era, many elements of these social democratic systems such as public health services and social housing programs remain in place. It could be argued that some of these countries have been strict about immigration and so are simply being good to their own people. However, actual immigration policies vary, and in recent decades Sweden, Great Britain, and Germany have welcomed large numbers of immigrants.

To provide a strong social safety net, these European countries plus Canada and Japan typically allocate between 25 and 32 percent of GDP to social services for all citizens, not just the poor.[15] The distinction is crucial—whereas in the US social safety net programs, such as they are, serve primarily the poor, in social democracies they serve everyone and thus achieve higher political buy-in across society.[16] Benefits include free or low-cost health care, large amounts of relatively affordable social (public) housing, free or inexpensive higher education, disability and unemployment assistance, generous parental leave, and childcare.

In social democratic countries the public sector has taken over certain sectors of economic activity deemed essential to meet human needs. Many businesses may also be nonprofit cooperatives. The public sector directly runs health care services in Britain, Spain, New Zealand, and most of Scandinavia. In other countries the public sector provides health insurance, and care is carried out by private providers. National governments also often own and operate train systems, phone companies, banks, and oil companies. Government ownership does not necessarily improve service to the public or solve environmental problems, and in parts of the world it has led to corruption. But in theory it can allow democratic control over sectors of the economy important for sustainability, access, and equity.

Democratic Socialism

Although it sounds similar to social democracy and exists on a continuum with it, democratic socialism is a system in which the public sector controls most or all economic sectors. Markets can still exist, but private accumulation is no longer the main motive for economic activity. Rather, people work for wages and to help meet community needs, and in turn democratically control the state. Within totalitarian socialism, in contrast, a self-perpetuating elite controls the economy and government. In practice the Soviet Union and the countries of eastern Europe did not do well either at meeting human needs or at protecting the environment.

If democratic socialism is highly decentralized, it verges on anarchism, a philosophy emphasizing grassroots self-governance through voluntary institutions with no formal state. In anarcho-syndicalism, workers control these institutions. Northern Spain and Catalonia serve as the world's leading example of this worker cooperative philosophy in action. Building on nineteenth-century cooperative traditions, workers collectivized many Spanish factories in the 1930s before being crushed by fascist leader Francisco Franco during the Spanish Civil War.[17] However, the northern Spanish city of Mondragon, discussed further below, still consists almost entirely of worker co-ops. Self-organized forms of grassroots socialist community are also prominent in Barcelona.

In democratic socialism businesses still operate within markets, allowing price signals rather than top-down mandates to guide production. But collective benefit rather than private accumulation is the purpose of economic activity. A long line of thinkers has endorsed versions of this approach. It has been argued that Martin Luther King Jr. was a democratic socialist, partly on the basis of a 1966 speech: "We are saying that something is wrong . . . with capitalism. . . . There must be better distribution of wealth and maybe America must move toward a democratic socialism. Call it what you may, call it democracy, or call it democratic socialism, but there must be a better distribution of wealth within this country for all of God's children."[18]

Bolivia's version of democratic socialism under President Evo Morales between 2006 and 2019 was inspirational for many. Through its emphasis on *buen vivir* (good living), the country sought an alternative set of social values to the neoliberal capitalism pushed by international development agencies, one based on indigenous traditions and basic human needs. Morales and his followers nationalized oil companies, redistributed land, built schools, provided pensions for those over sixty, and paid young people to stay in school. These actions helped reduce poverty in Bolivia from 36 percent to 17 percent between 2004 and 2017.[19] Although Morales in the end sought to cling to power too strongly, the country's achievements under his Movement toward Socialism (MAS) are impressive.

Others such as Rabbi Michael Lerner seek a more humanistic socialism emphasizing a transition of human values away from twentieth-century materialist economics and toward caring and compassion. Lerner's "spiritual socialism" draws from feminism, deep ecology, spiritual philosophies, and critiques of structural racism to argue for a political economy based on "the love of life and all beings."[20] It is interesting to think about the forms of education and incentives that might help bring about such a kinder, gentler political economy in the long run. Potentially public service requirements, using expanded programs such as the Peace Corps and AmeriCorps in the US, might help people understand those different from themselves and so lead to such a humanistic socialism.

OTHER TYPES OF SUSTAINABLE ECONOMY

Within these three basic paradigms, other economic approaches can help bring sustainability about on local or global levels. These changes relate to sustainability problems such as overconsumption, inequality, excessive financial speculation, and the scale of businesses (table 3).

A Degrowth Economy

Is continual growth in economic production and consumption sustainable? Or should we aim for stable or reduced production? Ever since the 1960s leading economists such as Kenneth Boulding, former president of the American Economic Association and the American Association for the Advancement of Science, have doubted that endless economic growth can take place on a finite planet.[21] Growth typically involves increases in resource consumption, pollution, and GHGs and is often associated with other negative externalities such as inequality and concentration of economic and political power. Indicators such as Gross Domestic Product (GDP) do not address environmental or social externalities, so evaluating economies purely on growth in production is very misleading in terms of sustainability. It is clear that GDP is not a metric associated with sustainability and needs to be replaced, particularly in light of the global climate crisis.

Optimists say continual growth is possible if production can be decoupled from resource consumption and externalities.[22] In this scenario economic activity would eventually be based entirely on reused materials and renewable energy. One strategy to bring about this transition, as Paul Hawken argues, would be to tax "bads" such as pollution and GHGs rather than "goods" such as products.[23] Others argue that completely decoupling economic growth in this way from resource consumption is impossible. Production in their view will always involve some negative externalities.[24] So the notion of continuous growth must be left behind.

"Degrowth" is an increasingly discussed alternative in which production and consumption would fall while economies shifted to renewable resources and energy as rapidly as possible. People would still build houses and buy products, but a stable or declining global population would live relatively simply with wealth spread more evenly. (Population

TABLE 3   STRATEGIES FOR SUSTAINABLE ECONOMIES

Three Paradigms	Description
Reformed capitalism	Regulate business to ensure environmental and social well-being; reduce capital's ability to influence government.
Social democracy	Regulate business; provide a strong social safety net; public takes over some economic sectors (e.g., health care, utilities).
Democratic socialism	Public controls all economic sectors, provides a strong social safety net, and allows markets to set prices.

Seven Emphases	
A degrowth economy	Emphasize growth in quality of life rather than material production; use pricing, education, and cultural change to reduce consumption.
A locally oriented economy	Limit corporate size and prevent monopolies or oligopolies; emphasize local/regional production and social/ecological accountability.
An egalitarian economy	Promote social equality and discourage private wealth accumulation through progressive taxation, high inheritance taxes, and limits on pay disparities.
A nonspeculative economy	Restrict and regulate speculative markets (e.g., derivatives, futures, stocks); institute high taxes on capital gains.
A triple-bottom-line economy	New corporate charters require environmental, social, and economic benefit; improved transparency and environmental/social impact reporting.
A cooperative economy	Promote or require worker, consumer, and producer cooperatives so as to maximize social benefit of economic activity.
A collaborative economy	The public sector works collaboratively with business, labor, NGOs, and other stakeholders to meet environmental, economic, and social goals.

City Strategies	
Invest in human capital.	Improve education, quality of life, and human services to stimulate new businesses and draw employers.
Promote Asset-Based Community Development (ABCD).	Bring stakeholders together to identify existing community strengths and build on them.
Develop business incubators.	Create spaces in which start-ups can have offices, share services, and receive technical support.
Embrace local hiring and contracting.	Require businesses to hire a percentage of workers locally and provide job training.
Undertake supportive physical planning.	Design neighborhoods, streets, and public spaces so as to support local businesses and revitalize neighborhoods around them.

NOTE: Elements of these alternatives can of course be mixed.

growth has already stabilized or begun declining in a number of industrialized countries and is likely to do so elsewhere if poverty decreases and the status of women improves.)[25] The degrowth concept has roots going back to the nineteenth-century economist John Stuart Mill, who believed society would reach a state with a constant population and stock of capital, in which people focused on human improvement and "the art of living."[26] The idea gained steam in the 1970s, perhaps as a result of the 1960s and '70s reaction against materialism, when Herman Daly proposed a steady-state economy as an alternative to the growth economy.[27] Criticizing "growthmania," Daly argued that the public sector should set minimum and maximum limits on wealth, stabilize population, and establish depletion quotas on resource use. Technology could still improve, but the emphasis would be on quality of life and sustainability of communities rather than the sheer quantity of goods and services produced.

Under degrowth, GDP would be replaced by new indicators of progress. For example, Daly and John Cobb have proposed an Index of Sustainable Economic Welfare (ISEW) that includes social and environmental welfare.[28] The country of Bhutan has created a "Gross National Happiness" Index that measures health, education, ecological diversity and resilience, living standards, governance, cultural diversity and resilience, living standards, community vitality, and time use. Such new metrics are important because they signal our collective priorities and values.

A Locally Oriented Economy

One of the leading alternatives to big-scale global capitalism is a locally oriented economy that sharply limits the size of corporations and encourages local or regional production for local markets. Potential benefits include increased competition, greater consumer choice, lower prices, a more resilient economy, and reduced transportation needs in certain industries. Smaller companies would be more rooted in local communities, with owners and managers living locally. As a result they might have more commitment to civic leadership and rarely would become "too big to fail." A decentralized economy might also have political advantages in that few corporations would get so big, and few industries so centralized, as to have undue political power. During the Covid-19 pandemic with the disruption of global supply chains, there was also recognition that more local production improves the resiliency of many businesses.

An obvious way to create this smaller-scale economy is to break up very large companies, especially if they have become monopolies or oligopolies. This has been done in the past at different times. In 1911 the US Supreme Court ruled that Standard Oil—the massive oil company founded by John D. Rockefeller in the nineteenth century—was an illegal monopoly under the 1890 Sherman Antitrust Act. The company was broken up into thirty-four smaller companies, some of which evolved into giants in their own right, such as ExxonMobil and Chevron. In 1974 the US Justice Department filed suit against AT&T, at that time the world's largest telephone company, on similar grounds, resulting in its 1984 breakup into seven regional phone companies. (Parts of AT&T have since come back together.)

Another way to decentralize economies would be to emphasize what Michael Shuman calls the "community corporation"—organizations of any type that are owned locally, emphasize meeting human needs, and are anchored in particular cities.[29] Usually nonprofit, these often receive assistance or preferential contracting opportunities from local government. The community-owned model is the opposite of the footloose global corporation, searching for the lowest costs regardless of the social or environmental consequences. In recent generations such companies have been able to play cities off against one another to secure enormous tax breaks for new facilities, often only to leave town a few years later when cheaper labor or better incentives in another part of the world appeared.

A range of tools to decentralize financial services can help jump-start local economies by making financing available to people and small businesses who do not otherwise have access to them.[30] Credit unions (nonprofit, customer-owned banks) are one of these. Microfinance institutions—small-scale, decentralized banks making small loans to low-income individuals who would not otherwise qualify for them— represent another. Best known through developing-world examples such as the Grameen Bank in Bangladesh, microfinance institutions exist in developed nations too. Finally, the public sector can require that larger, commercial banks engage in community development lending in places where they do business.

An Egalitarian Economy

A related ideal is an egalitarian economy—a political and economic system that maximizes social equality by limiting private capital accumulation. The means to do this are relatively simple and will be discussed at greater length in chapter 4. They start with highly progressive taxation—a tax structure with marginal tax rates increasing the more one earns, up to, say, rates of 90 percent or more above $5 million in income or capital gains each year. Progressive taxation could exist for businesses as well as individuals. Other possible tools include a high minimum wage (enough for a family to live on), a guaranteed annual income, and/or requirements that the top earner in a company make no more than, say, ten times as much as the lowest-paid worker. These measures to reduce economic inequality would have ripple effects, making it possible, for example, for working-class people to afford healthy, organic food and housing near their jobs.

A Nonspeculative Economy

Another sustainability-oriented economic strategy would be to reduce, discourage, or eliminate financial speculation. Speculation tends to promote inequality, since it offers some people a way to get rich quickly while raising prices of housing, land, and consumer goods for others. It also reinforces antisocial values of greed and selfishness. Unrestrained speculation can jeopardize the entire world economy, as we saw with the 2008 Great Recession.[31]

Speculation has long been part of market economies, illustrated for example by the Dutch tulip bubble of the 1630s, in which speculators paid the equivalent of a year's wages for rare bulbs of what was then seen as a luxury flower, only to have the market crash. Housing bubbles have been found historically in many cities and countries,[32] and they can have serious impacts on household welfare, first by making housing unaffordable for many and then by wiping out a main type of household equity when the bubbles burst.

Following economic deregulation in many countries in the 1980s and 1990s, financial speculation mushroomed. The FIRE (Finance, Insurance, and Real Estate) sector of economies—which encompasses many speculative activities—grew rapidly as a result. Journalist Susan Strange has used the term *casino capitalism* to refer to the enormous growth under neoliberal policies of financial markets only loosely under democratic control.[33] As the "house," the financial services industry itself is the biggest winner in the casino economy. Gambling came to pervade many economies quite literally, with lotteries and casinos spreading. Perhaps it was fitting that Donald Trump owned a casino in Atlantic City and also dealt primarily in real estate development, frequently a speculative profession, before becoming president of the United States.

Far from being an isolated case, he represents values that have come to the fore in many societies.

Because of the country's weak social safety net and lack of pensions (guaranteed retirement income from employers), many US citizens are forced to become speculators because their retirement savings accounts are invested in the stock market and much of their other wealth is in real estate (i.e., their home). Thus citizens end up supporting corporate America's desires to cut environmental regulations and worker protections because they want to keep stock prices high.

There has been a long history of the public sector limiting speculative economic activity. The US Glass-Steagall Act of 1933 (repealed in 1999 by the Gramm-Leach Act) prohibited banks from dealing in risky securities. More recently the Dodd-Frank Wall Street Reform and Consumer Protection Act of 2010 attempted to limit speculation by banks and within futures markets. Dodd-Frank was eviscerated during the Trump administration. On a smaller scale, some cities have adopted antiflipping taxes to discourage short-term real estate speculation and subsequent inflation of housing prices. The Province of British Columbia does this by taxing profits from short-term real estate sales at 100 percent as business income, rather than at the 50 percent capital gains rate that a longtime homeowner would pay upon selling their house.[34]

How much speculative activity is really needed within a healthy capitalist economy? Arguably little or none. Public banks and credit unions could replace for-profit banks altogether, providing investment capital to businesses and ensuring equitable access to capital.[35] Many financial services, including much investment banking and tax advising, exist only because legislators allied with the wealthy have allowed complex financial instruments ripe for exploitation to come into existence. Simpler, fairer, and less regressive tax and banking systems wouldn't need such elaborate edifices of financial advising and trading. Likewise, large insurance industries exist only because of social insecurity, including the lack of public health care in the US. Removing the need for insurance by having a government-funded retirement system that guarantees benefits could save enormous sums—freeing up capital to meet real human needs.

High capital gains taxes (taxing investment income rather than salaries) would have the benefit of discouraging speculation throughout the economy. US Republicans pushed for and won low capital gains rates in the 1990s and 2000s, but these simply encouraged speculation and meant that the wealthy (whose wealth comes more from investments than salaries) were taxed more lightly than the middle class (who are taxed on their wages). At the level of city government, strong land use controls would disincentivize the speculative holding of vacant land in expectation of development windfalls.

Perhaps the most revolutionary antispeculative move would be a small global tax on financial transactions. The resulting funds might then be reallocated to poverty remediation. In 2010 a coalition of NGOs based in the United Kingdom proposed such a global "Robin Hood tax," also called a "Tobin tax" after economist James Tobin, who suggested it. This tiny tax of 0.05–0.5 percent on sales of all stocks, bonds, commodities, derivatives, or other financial instruments would raise enormous sums that could then be used to fight poverty and climate change. A much more limited financial transactions tax considered by the European Union in the early 2010s promised to generate €57 billion per year.[36]

A Triple-Bottom-Line Economy

Yet another way to bring about more sustainable economies would be to apply the triple-bottom-line principle. Under this, corporations would be issued new charters requiring them to meet social and environmental goals, in addition to seeking profit. Benefit corporations (discussed below) are already chartered in this way.

Today corporate licensing is weak in virtually every country. For two centuries now corporations in the US have operated under charters granted by individual states competing to see who can offer the weakest licensing requirements. Delaware has long been the winner in this "race to the bottom" with what is known as the "Delaware Loophole," which allows companies to avoid paying high tax rates.[37] As a result, two-thirds of Fortune 500 companies and 1.2 million US businesses overall are chartered in that tiny, little-known state.[38] Many also shelter some of their revenue from taxation by pretending to be based in overseas locales such as the Cayman Islands.

A logical remedy would be for the federal government to step in and require national licensing of corporations. Ralph Nader, Mark Green, and Joel Seligman proposed this approach in 1977 in their aptly titled book *Taming the Giant Corporation*.[39] In 2018, Senator Elizabeth Warren introduced a bill—the Accountable Capitalism Act—to likewise require large corporations to get a federal charter. Under Warren's bill, companies would acknowledge that their responsibilities go beyond maximizing return to shareholders. They would need to give workers

40 percent of board seats, would need three-quarters of the board and shareholders to approve any use of company funds for political purposes, and would require top executives to hold stock for five years before selling it (to decrease incentives for short-term profiteering). More representation of women and people of color on boards could be required as well. A 2020 California law requires companies based in the state to have two or three individuals (depending on board size) from underrepresented population groups on their boards by 2022.[40]

Requiring companies to prepare social and environmental impact statements on major actions or their annual operations would also help enforce a triple-bottom-line approach. Companies already have to prepare environmental impact reports on the construction of new facilities in the state of California, and such reporting could be expanded into more holistic sustainability and equity analysis.

A further way to emphasize the triple-bottom-line approach is through benefit corporations. In some US states, Italy, Colombia, and a growing number of other countries, new corporate licensing options allow businesses to be created with a mission to have positive effects on society and the environment as well as to make money. Usually they are required to prepare annual reports on these benefits. Third-party organizations can certify benefit corporations through mechanisms such as the "B corporation" credential now held by more than 2,600 companies in sixty countries. City governments can prioritize assistance for this new form of capitalism, while state governments can make sure that their business licensing procedures facilitate it.

A Cooperative Economy

Yet another type of sustainable economy would emphasize cooperative businesses. Within this model, businesses would be limited to co-ops collectively owned by worker, producer, consumer, or local resident groups. Although it sounds idealistic, this form of enterprise is already common worldwide. Although cooperatives often operate quietly without calling attention to themselves, they could be more vigorously supported by cities and towns and made a cornerstone of local economic development.

Cooperatives may be either nonprofit or for-profit but are likely to have a socially responsible, public service orientation due to their collective ownership. Food co-ops—grocery stores owned by their customers—are perhaps the best-known examples of this form of business. Typically these emphasize healthy, organic, locally produced foods and give their member-owners annual rebates based on their volume of purchases. Some of these co-ops are quite large and sell household products in addition to food. Co-op Danmark, for example, is the largest retailer of consumer goods in Denmark, with 1.4 million members and a 40 percent market share.[41] Other forms of consumer co-ops are active in many sectors of the Japanese economy, with more than seventeen million members.[42]

Producer co-ops market and distribute goods from associations of farmers or manufacturers. American producer co-ops include large businesses such as Organic Valley dairy products, Sun-Kist oranges, and Land-o-Lakes butter. They play an important role in countering the small handful of for-profit companies that exercise oligopoly power when purchasing food from farmers. Encouraging producer co-ops and distribution hubs is one way to help local food systems move toward sustainability and equity.

Credit unions are nonprofit banks owned by their members. They typically have a stronger orientation toward serving local communities than commercial banks and have used the slogan "Serve people, not profit." Some eighty-five thousand exist worldwide, serving 237 million members, with six thousand in the US.[43] In the Middle East and Latin America, rotating savings and credit associations (ROSCAs) have been operating informally for centuries, providing small-scale financing as well as mutual support within groups of local residents. However, most are relatively small. Accumulating savings and credit associations (ASCAs) are similar but keep an internal fund and develop a surplus, which is returned to members.

Worker co-ops—in which workers own and control the business—are found worldwide. Historically one of the best examples has been Mondragon, in northern Spain, where since the 1950s an economy has evolved composed of 266 different worker-run businesses employing eighty-one thousand people.[44] Workers must buy into the Mondragon cooperative to start with but then have a say in its management. During economic downtimes the network may move workers between cooperatives to maintain full employment, or if necessary all workers take reduced hours rather than having people laid off. Over more than seventy years this cooperative has done surprisingly well despite the pressures of a competitive global economy.[45] The northern Italian region of Emilia-Romagna is also a prime example of a place where networks of co-ops have created a strong and creative economy, generating one-third of that region's economic output.

A Collaborative Economy

In a collaborative economy, a type of political economy common within social democracy, the public sector operates as a main coordinator of economic activity and investor. It maximizes social and environmental well-being by working collaboratively with business, labor, and other stakeholders. These players commit to a multistakeholder policy process. This alternative was the norm for much of the twentieth century in Sweden and other Nordic countries. Beginning with the rise of the Social Democrats in the 1930s, different Swedish political parties had the same general goals of modernizing the country, building a strong social safety net, valuing the role of workers and labor unions, and protecting the environment.[46] Not all these things were done perfectly. But politicians, corporations, and labor unions often agreed on policy aims and respected one another. The tone of political discourse was strongly collaborative and deliberative, rather than partisan and political.

Somewhat similar conditions existed in the Netherlands during the second half of the twentieth century, especially around environmental policy. In Japan government, business, and labor also collaborated closely. Unions went out of their way to avoid strikes and in return were rewarded with the world's most generous pension system.[47] Such collaboration can have downsides, certainly—Japanese unions, for example, have been accused of being too cozy with business. But the upside includes the ability for societies to respond powerfully, creatively, and in relatively unified fashion to the challenges of the time.

How might a collaborative economy come about? One requirement is a strong ethic of civility and shared purpose. Such a situation seems far away in the US, Great Britain, and other countries in which political polarization has increased. However, committed and ethical leadership can make a difference by calling attention to common goals and needs. The urgent need to prevent pandemics, address the climate crisis, and promote social and racial justice may form the political foundation for a more collaborative approach. Forums can be established in which competing interests can build trust. The media can play a role in shining a light on serious abuses of economic power and potential strategies to address them rather than dwelling on the minor horse races of political life.

In the past, crises such as war or the Great Depression have brought political factions together, resulting in a more collaborative and unified society. Whether slower-moving disasters such as climate change, pandemics, and growing social inequality can have the same effect remains to be seen. But leaders can and must seek to use these motivating factors to bring about more constructive processes that aim to redress inequity and promote sustainability.

CITY-SCALE SUSTAINABLE ECONOMY STRATEGIES

Even while we collectively rethink large-scale economic systems, creative local economic development policies are needed in cities and towns. Local economic development is a difficult challenge in many places. What can be done when there aren't enough jobs for people? When storefronts are vacant? When existing companies close up or leave town? When the local tax base isn't enough to support needed services? Twentieth-century "public choice" theory would argue that places become obsolete as the economy changes and so people need to leave them. However, these places are people's homes and have important cultural, historical, and social value. Thinking that it is acceptable to leave places behind and keep moving on is no way to create sustainable communities. We need to imagine equitable transitions.

Conventional economic development strategies often have worked against sustainable cities. One main strategy has been to use public subsidies including tax breaks and publicly funded infrastructure to lure new corporate employers from out of town. The results are underwhelming. Local incentives typically don't result in more jobs or tax revenue, despite their high cost,[48] and newly arrived businesses may themselves soon close or leave town.

Zoning large amounts of land on the edge of the city for new business parks or malls is another typical local strategy, often coupled with tax incentives and publicly subsidized roads, water, and sewers. If successful, these facilities may kill businesses in older parts of the city and promote suburban sprawl with increased motor vehicle use. Retailers such as Walmart often pay minimum-wage salaries with few benefits, which local residents can barely live on. Although it may benefit from development fees initially, the local government in the long run finds itself faced with higher costs for repaving roads, fixing water and sewer systems, and otherwise maintaining the infrastructure of suburban sprawl.[49] A more sustainable path of development would be to incrementally redevelop buildings within the existing urban area for new businesses, using already-built infrastructure and putting jobs and stores near where people already live.

Trying to keep alive old, extractive industries such as mining, logging, or oil drilling doesn't help either. Often environmental regulations are weakened or waived to do so. The local community may enter a boom-and-bust cycle, with periods of "success" drawing workers from far and near, only to have jobs dry up a few years later. These extractive industries are often fundamentally unsustainable and may fade away no matter what assistance is offered them. The western US has seen this happen in many places, as lumber and mining industries tend to be cyclical and in long-term decline.

When it comes to urban planning, first do no harm, as the medical profession has long counseled itself. There's usually no magic bullet to fix a challenging local economic situation. Rather, what's needed is a slow rebuilding of civic health through many small, thoughtful steps (table 3). Political action for structural change may also be necessary, for example to change the way schools and local governments are funded so that they are not so dependent on the local property tax and so that resources are shared between wealthy and struggling communities. Solidarity with surrounding jurisdictions can be important so that one city doesn't try to steal tax base or employers from its neighbors and instead both undertake synergistic economic development strategies. Coordination with higher levels of government is essential. Often states, regions, provinces, or national governments can offer technical support, grants, or infrastructure that can make local initiatives possible.

Invest in Human Capital

One main approach likely to lead to sustainable economic development is to spend available funding on education, social services, culture, affordable housing, and civic amenities rather than tax subsidies for businesses. These investments build human capital—strong, capable, and secure local residents who can build businesses and attract external employers looking for a good workforce. Such steps also improve quality of life, attracting talented new residents who can in turn help develop the local economy.

Where funding for such quality-of-life improvements would come from will vary by context. To some extent it is a question of local priorities. But following decades of tax cutting, US local governments are often very limited in their ability to raise revenue, and states are legally required to balance their budgets each year and thus are limited as well.

So much larger federal support for cities and towns may be needed. Taxing capital gains (i.e., profits from investments) at the same rate as ordinary income—a step that would greatly promote social equity—would generate upwards of $192 billion annually.[50] Given a US population of 332 million, the federal government could then provide an annual grant of about $580 per person to local governments. For a town of one hundred thousand people, that works out to $58 million. Other, more local sources of funding are possible as well. The process known as redevelopment in some states allows cities to identify neighborhoods in need of improvement and then issue bonds to be paid back with increased tax dollars after new urban design improvements are made, a tactic known as tax-increment financing.

Apart from financial investment, Richard Florida's theory of the "creative class" suggests that local policies of social tolerance combined with high quality of life can promote a concentration of creative individuals, who in turn contribute to economic development.[51] Celebrating demographic diversity, artists, LGBTQ communities, and local cultural traditions can thus be itself an economic development strategy. This perspective has been criticized for being elitist in that it favors young, well-educated knowledge workers over older working-class communities and may promote gentrification of older neighborhoods and result in displacement. The link between the arts and job development is also by no means certain. However, it remains highly influential within local economic development planning, and if combined with protections for existing renters and concerns about social and racial justice it can be part of a human-capital-and-quality-of-life strategy.

Promote Asset-Based Community Development (ABCD)

A somewhat different approach is that of Asset-Based Community Development (ABCD). Developed by John L. McKnight and John P. Kretzmann in the 1990s, this philosophy calls for identifying the existing strengths of a community and building on them.[52] Instead of trying to meet perceived community needs through outside interventions, organizers work within neighborhoods to empower individuals around existing assets. These assets are often mapped in creative ways. Organizers identify local skills, institutions, networks, associations, cultural history, and physical facilities. Community members brainstorm ways to build on existing strengths.

Although criticized for potentially downplaying the need for external assistance, the ABCD approach is often seen as a good starting point for bottom-up local economic development. It builds on strengths of communities and empowers residents to be agents of change. The city of Hamilton, Ontario, for example, used an asset-based approach to develop a 2011 Neighborhood Action Strategy focusing on eleven neighborhoods. The program worked to connect youth with employment and volunteer opportunities, added on-site services and resident engagement activities to social housing projects, and created an urban farm at which low-income residents develop work skills and receive free food.[53]

Develop Business Incubators

Since being pioneered in the 1980s, business incubators represent another popular local economic development strategy. More than seven thousand are in existence worldwide. These buildings nurture start-up businesses, providing affordable rent, access to technology and services, access to loans and investors, and assistance with financial management, accounting, marketing, and business development. Fledgling companies also learn from one another. Incubators may be either privately or publicly owned and are often focused on technology companies and/or associated with universities. One leading example is Boulder, Colorado's Techstars accelerator, which since 2007 has helped more than 150 fledgling companies connect with mentors and raised more than $1 billion in start-up funding.[54]

The effectiveness of business incubators depends on rigorous assessment of potential tenants, good management, strong sponsors, good connection to services and networks, physical facilities that promote interaction, and government policy that will support businesses as they emerge.[55] For sustainability purposes, incubators can focus on environmentally responsible companies and/or disadvantaged local populations. These facilities represent a bottom-up economic development approach far preferable to chasing out-of-town industries through public sector giveaways.

Embrace Local Hiring and Contracting

Local hiring policies require businesses to hire a certain percentage of local residents for jobs and may also require businesses to hire women and people of color. The advantage for the community is that any increase in employment goes to existing residents rather than drawing out-of-town workers who might then increase traffic, pollution, and demands on local services. Frequently local hiring policies arise from development agreements between cities and large companies that want to locate there or expand existing operations and need permit approvals from the local government in order to do so. These agreements may also include training programs to help low-income residents gain the skills they need for the new jobs.

One of the most dramatic examples of a local hiring policy was the City of San Francisco's 2010 ordinance, which required 50 percent of construction workers employed by projects after 2017 to live in the city.[56] Complying with such policy can of course be difficult if a jurisdiction has little affordable housing, so other policies need to ensure that this exists as well.

Similarly, local contracting policies can require public agencies and any firm doing business with the local government to contract with local suppliers for all their needs. For example, the city council of Preston, England, a struggling small city in Lancashire, persuaded six main public agencies to spend their money in the city or county starting in the early 2010s. Whereas in 2013 these agencies spent £330 million of their budgets locally, by 2017 the total had risen to £597 million, despite overall budget shrinkage. Officials often had to break large public contracts up into small pieces so that small local businesses were able to bid on them. But the result was broad-based revitalization of a moribund local economy.[57]

Undertake Supportive Physical Planning

Local economic development efforts often focus on creating physical space for new businesses by rezoning land for new business parks at the urban fringe, thus fueling suburban sprawl. However, cities can instead seek ways to revitalize older, more central sites for economic development. This path of physical development can be hard—old buildings must be rehabilitated, land-remediated neighbors consulted with, expensive multistory mixed-use structures designed instead of one-story cookie-cutter boxes, and traffic and parking issues dealt with. Cities will need greater staff capacity in order to do the planning and coordination needed to help such development happen. Urban land is also typically more costly than sites at the fringe. Hence the need at higher levels of government to rethink the economic incentives that keep sprawl development artificially cheap by leaving out its social and environmental costs, and the need for local governments to restrict sprawl development through regulation in the absence of good economic signals.

Asheville, North Carolina, is an example of a small city that rehabilitated historic downtown buildings and spaces for new businesses rather than razing them and constructing a giant mall and convention center, as proposed in the late twentieth century. By working with its historic, fine-grained urban fabric, the city created an arts district with more than 170 galleries and studios as well as an attractive, human-scaled city center that draws thousands of tourists. Blocks that were shabby and half-abandoned are now full of life. A form-based code (a revised and simplified type of zoning focusing on building form and character) has helped guide development. Revitalization has been led by many small players (architects, property owners, nonprofits, foundations, and politicians) rather than a few big agencies or developers.[58]

CONCLUSION

The question of how a sustainable economic foundation for sustainable cities might come about is an enormous one, requiring us to rethink philosophies at multiple scales. Societies will need to consider big-picture choices between reformed capitalism, social democracy, and democratic socialism. They will also need to rethink "growth" and consider new economic development philosophies focusing on social equity, cooperative businesses, limits to speculation, increased collaboration between stakeholders, triple-bottom-line analysis, and smaller-scale and more local businesses. New indicators of progress will be necessary. Negative externalities of economic development such as pollution and climate change will need to be better incorporated into business and public sector decision-making. Then many local economic development tools will need to be employed.

In sustainable cities of the future, businesses—likely a mix of for-profit and nonprofit—will prioritize clean methods of production rather than polluting ones. They will provide living wages for all residents and will emphasize the meaning and dignity of work. Teachers and nurses will be compensated at the same level as bankers managing other people's money. Workers will have the opportunity to help run their workplaces, and community members will have the opportunity to collaborate with business owners when their interests overlap. Stronger social safety nets will ensure that nobody will need to worry about suddenly losing all their income or health insurance if their employment situation changes. Businesses will be more strongly tied to the cities they operate in, won't be able to manipulate government or the media against the common good, and will help build the long-term welfare of communities and regions as well as provide truly needed goods and services. Across many scales of action and levels of government, capitalism will have been either placed within a stronger framework of public control or replaced by economic systems that emphasize social and ecological welfare rather than individual profit.

4

How Can We Make Affordable, Inclusive, and Equitable Cities?

Jasmine was kicked out of her Australian family's home when she was eighteen. She spent the next three years sleeping in cars and on friends' couches before giving birth to her daughter at twenty-one. Determined not to be homeless with an infant, she found space in a shelter but was assaulted by a male staff member and had to leave. She then applied for public housing. However, in New South Wales there were fifty-two thousand individuals on the list ahead of her. The average wait time was more than ten years. To keep a roof over her head she entered one abusive relationship after another. Finally at the age of thirty-one she was able to obtain her own public housing apartment and a measure of stability. She and her daughter faced years of recovery from their lives of trauma.[1]

Anwar emigrated with his family from Egypt to Toronto in the early 2010s. As with many immigrants to First World countries, the only housing he could find was in an aging high-rise building with hundreds of other immigrants in the Rexdale complex in a declining inner suburb. It was a home with many problems. The building's elevators, washing machines, and air conditioners often failed. Vandals spray-painted graffiti on the walls. Burglaries, drug dealing, and racial taunts were common. Anwar and his family had great difficulty feeling welcome in their new community.[2]

George graduated from high school after growing up in a bleak public housing project known as "the Bricks" in one of Houston's poorest neighborhoods.[3] Like many athletically inclined African American young men denied access to other opportunities, he dreamed about becoming a professional athlete. When that option didn't pan out, he started using drugs, and an arrest for a ten-dollar drug deal put him in state prison for ten months. Afterwards he and a friend decided to move north to Minneapolis in hopes that job opportunities would be better. After working as a bar bouncer and security guard for the Salvation Army, where he earned a reputation for kindness to other employees, he was laid off in the Covid-19 pandemic. After a store clerk called 911 on suspicion that George had given him a counterfeit twenty-dollar bill, police officer Derek Michael Chauvin arrested him, handcuffed him, and killed him by kneeling on his neck for eight minutes and forty-six seconds. As with many police killings of Black men, this would normally have been the end of the story, except that a seventeen-year-old girl took a video of the event with her cellphone and posted it online, which led to a massive protest movement around the world demanding recognition of racial injustice and an end to systemic racism.

Such stories, some more dramatic than others, illustrate the social and racial equity struggles that many urban residents face across the globe. Because of poverty, structural racism, lack of economic opportunity, and various forms of discrimination, millions struggle for decent living conditions and life chances.

Let us imagine instead cities that are humane, affordable, inclusive, and equitable for all: places where opportunistic politicians no longer stoke hatred and fear; where we recognize past injustices and actively seek to address them; where safe and affordable housing for everyone is a priority; where good schools and access to clean air do not depend on your zip code; and where ending poverty and racial inequality is a focus of public policy, and resources are reallocated for this purpose from prisons, the military, and the police.

Social and racial equity needs to join the climate crisis at the forefront of planning for sustainable cities. Too often sustainability has been associated mainly with urban greening. Critics argue that sustainable city efforts lead to "green gentrification" when new parks, bike lanes, and transit lines draw well-off newcomers into urban neighborhoods, displacing existing residents.[4] Instead of emphasizing primarily environmental improvements, sustainable city visions need to prioritize improvements in social equity and justice, while reducing divisions among people and jurisdictions. They need to create what Julian Agyeman and others have called "just sustainabilities."[5]

THE CHALLENGES AND OPPORTUNITIES OF DIVERSITY

Because of their population size, variety of economic niches, and role as way stations, cities have long been more diverse and tolerant than rural areas. "City air makes you free" was the medieval German saying, probably arising from the legal principle that former serfs were considered free after a year's urban residence.[6] Racial and religious minorities were often tolerated only in cities. Cities have also been critical for women's rights historically and have often featured lower birth rates because female residents have increased access to family planning, education, and economic opportunity.[7] LGBTQ people have flocked to urban areas and have continued to do so until recently, when somewhat greater acceptance throughout societies has increased their safety in rural areas. Urban residents have not always been tolerant of diversity, of course, but cities have been the safer place to be for all these groups.

Urban diversity has continued to grow in recent times because of massive movements of people around the globe. This mixing of peoples is much of what makes urban life so dynamic and interesting. In the 2010s more than 31 percent of Londoners were born outside Great Britain and 55 percent identified as "other than white British."[8] In 2018, 40 percent of New York City's population was foreign born.[9] Sao Paolo's eleven million people come from more than one hundred different ethnic groups originating in Europe, Africa, the Middle East, and Asia. Given the scale of the expected climate migrations, we can expect that cities will become even more diverse in the future and so will need to plan for equitable inclusion.

The populist backlash worldwide has arisen partially because of this growing social diversity (caused in part by pressures of global warming). Yet at the same time that intolerance seems on the rise, Black Lives Matter has become the largest US social movement since the 1960s and a substantial force globally (figure 6). This movement, which seeks tolerance and full racial equality, has forced many white Americans to acknowledge the systematic racism that has formed the African American experience. Similar protests against racism, intolerance, and police violence have occurred in countries such as Britain, France, Belgium, Australia, and New Zealand.[10] A new generation is calling for greater acceptance of diversity and action to create inclusivity.

Some of the world's most successful examples of urban sustainability currently are unfortunately not very diverse. Singapore is often held up as one of these, despite having relatively little diversity (with a primarily Chinese and Malay population) and possessing an authoritarian

FIGURE 6. Black Lives Matter protests have been held worldwide, for example here in Brisbane, Australia. (Andrew Mercer/Creative Commons.)

government. Portland, Oregon, often seen as a best-practice example of American urban planning, is a relatively white city, although growing less so over time. Homogenous societies are by definition more unified, and a strong argument can be made that they can thus more easily come together around new policy initiatives and develop political support for steps to enhance the common welfare.

However, increasing diversity is a fact of life within cities and societies worldwide and offers many benefits; there is no choice but to embrace it. Discrimination, exclusion, and conflict present costs to societies and are just wrong. Building walls to keep out migrants is unlikely to work. Many members of dominant populations—those who are poor, immigrants, gay, female, nonbinary, transgender, differently abled, elderly, or of different religious persuasion—are "others" as well. What is clear is that sustainable cities need to accommodate everyone, celebrate diversity, and draw strength from difference.

STRATEGIES FOR AFFORDABLE HOUSING

A starting point for an inclusive city is for people of all types to be able to afford to live in it. Most of the world's nations agreed in 1976 at the first United Nations Conference on Human Settlements held in Vancouver, B.C., known as Habitat 1, that "adequate shelter and services are a basic human right" and that "the use and tenure of land should be subject to public control."[11] Housing as a human right was reaffirmed more recently in Sustainable Development Goal 11. However, this human right has yet to be guaranteed in most parts of the world.

Within industrial societies, growing numbers of people are homeless or live in substandard or overcrowded housing. The norm for affordability used in the US is that families should not pay more than 30 percent of their gross income on housing, maintenance, and utilities. However, much of the public has no choice but to do this. Researchers at Harvard's Joint Center for Housing Studies have put the number of US "cost-burdened" households at 38.9 million.[12] According to the National Low-Income Housing Coalition, the US has a shortage of about seven million affordable homes.[13] Similar shortages exist elsewhere. Britain, for example, is estimated to lack nearly four million homes.[14]

Some countries have actively tried to house everyone living within their borders. The Swedish government constructed one million new residences between 1965 and 1974 for a country of eight million people, in large part by funding local nonprofit Municipal Housing Corporations to build affordable rental units.[15] British county councils and other agencies built the majority of rental housing in the UK during the second half of the twentieth century. The French government has had a national goal of 20 percent social (affordable) housing,[16] and in 2015 it passed a law restricting rents in large cities to no more than 20 percent above the median for similarly sized and located units.[17] The US also funded its cities to build large amounts of public housing in the postwar period, but this was often designed to fail—built as cheaply as possible within large modernist buildings that segregated and isolated the poor without linkage to the range of social services, education, and structural reforms needed to holistically address poverty.[18] As a result, much US public housing became uninhabitable and has subsequently been torn down.

In developing countries the amount of housing constructed by the public sector has been smaller, and some one billion people live in informal settlements—self-built housing that is often woefully substandard, without running water or sanitation.[19] The challenge now is to upgrade these settlements by providing residents with legal title to land, better infrastructure (roads, sewers, water, electricity, and communications), services (education, police and fire protection, job training, health care, and social services), and sustainable economic opportunities.

Could decent housing be provided to everyone in the world? Some would say the cost would be prohibitive. However, the approximately $6 trillion that the US spent on its early twenty-first-century wars in Iraq and Afghanistan could have built approximately three hundred million housing units for poorly housed people in the developing world, assuming average costs of $20,000 for simple dwellings, providing homes to some one billion people.[20] Even at North American construction costs, that money could have built more than one hundred times the housing needed to house the approximately 568,000 people who were homeless in the United States in 2019.[21] So as usual with sustainability topics, the question is one of priorities, political will, international cooperation, values, and commitment rather than resources (table 4).

Build More Housing

The simplest strategy for providing affordable housing to everyone is to build lots more housing throughout metropolitan regions. Sheer volume can at times bring prices down. Having more new, good-quality, energy-efficient homes also makes it likely that some people will be able to leave older, less adequate housing behind and that the quality and energy efficiency of the overall residential stock will be improved. New housing should not convert greenfields into suburban sprawl; rather, it can reuse urban infill sites to create compact new neighborhoods built around existing infrastructure and transportation systems.

Housing has been underbuilt in the US and other countries in recent years in part because of local political opposition (often known as NIMBYism, for Not-In-My-Backyard) and lack of available, appropriately zoned land.[22] To get around these problems, the YIMBY movement—which stands for Yes-In-My-Backyard—has argued for eliminating single-family zoning, upzoning other areas for higher densities, and streamlining environmental review to enable rapid building. Advocates have also supported creation of municipal housing trust funds and state-level funding programs to provide grants to nonprofits to build affordable housing.

To ensure that housing is available to people of all income levels and backgrounds, state or regional agencies can require that all cities and towns zone for their "fair share" of housing affordable to households at different income levels. Then higher-level governments can use incentives and mandates to make sure this happens. On the "carrot" side they can provide funding for affordable housing and related planning.

TABLE 4 STRATEGIES FOR AFFORDABLE HOUSING, INCLUSION,
AND SOCIAL EQUITY

Affordable Housing	Description
Build more housing.	Remove barriers to increasing the quantity of housing in existing urbanized areas.
Expand social housing.	Promote forms of housing out of the speculative market (publicly built housing; housing built by NGOs; cooperative housing; limited-equity housing; community land trusts); limit speculative gains on private sector housing.
Expand rent controls, subsidies, and tenant protections.	Restrict rent increases, subsidize rents for low-income tenants, and protect renters from displacement.
Adopt inclusionary zoning.	Require private sector developers to integrate a certain percentage of affordable units into market-rate projects.
Increase lower-half incomes.	Raise minimum wage and adopt more progressive tax structures so as to ensure everyone has enough income to afford housing.

Inclusion	
Change institutions.	Rethink policing, the courts, the prison system, health care, planning, municipal agencies, education, and other social institutions so as to remove bias.
Enact affirmative action and reparation policies.	Proactively give opportunities to members of underrepresented and disadvantaged groups; approve special programs to address past injustices.
Treat migrants kindly.	Provide immigrants with social services, housing, ability to work, and pathways to citizenship; honor migrant cultures; protect migrants from deportation.

Equity	
Adopt more progressive tax policies.	Ensure steep increases in marginal tax rates for higher levels of income; tax capital gains the same as income; provide guaranteed income for the poor; raise wealth and inheritance taxes.
Ensure spatial equity.	Ensure that all people have access to all neighborhoods; equalize schools, services, and public amenities between neighborhoods and jurisdictions; share revenue across jurisdictions.
Protect vulnerable residents from displacement.	Protect existing residents and cultures as neighborhoods change through eviction controls, rent subsidies, and one-for-one replacement of housing units.
Retrain ourselves.	Emphasize understanding of historic injustices and actions; address these within education and organizational processes.

On the "stick" side they can allow affordable housing developers to override local regulations, or condition local receipt of infrastructure funding on compliance with fair share affordable housing goals. Massachusetts's Chapter 40B regulation, which allows developers to override local zoning when local affordable housing thresholds are not met, has helped create almost seventy thousand affordable units since 1969 (and more than 40 percent of them are for families making less than 80 percent of county area median income [AMI]).[23]

This "sheer volume" strategy has some problems, however. If the for-profit development industry continues to dominate housing construction, it may simply focus on providing units to the middle and upper classes where more profit is to be made. Newer, high-quality housing units will not necessarily trickle down to the poor in any reasonable time frame. New residences may also not be located near jobs or high-quality schools, or in the communities that most need them. So housing policy also needs to emphasize equity and appropriate location.

Expand Social Housing

A more far-reaching solution is to create a large amount of housing that is permanently removed from the speculative market, as some of the countries of northern Europe have done. Giving the unfortunate history of public housing in the US, the American public sector rarely constructs housing itself anymore. However, it helps nonprofit organizations build affordable housing by offering grants, loans, technical assistance, and tax credits to investors (notably, the Low-Income Housing Tax Credit or LIHTC). Thus a small US social housing sector is slowly emerging. These groups are often highly responsive to local communities and frequently provide supportive services for low-income individuals along with well-designed, energy-efficient residences.[24] Federal, state, or local action could accelerate the emergence of this social housing sector, primarily through funding. A number of cities have already set up revolving loan funds to support them. Denver, for example, created its Affordable Housing Loan Fund by increasing property taxes on houses valued at more than $500,000 and levying development impact fees, methods that raise some $16 million annually. In addition to supporting nonprofit builders, the city is using this fund to buy vacant high-end apartments and turn them into affordable units.[25]

Social housing programs elsewhere have fared better than US public housing because units are often intended for households of many income levels, are of higher quality, provide better mixes of unit types and sizes, and constitute a larger share of total housing. Great Britain began a social housing program in 1865, offering units in detached, semidetached, townhouse, and apartment forms. In Sweden's Million Home program, the government mixed apartment and detached housing within transit-oriented developments outside city centers and made these available to renters or to buyers through co-op arrangements with price controls so as to remain affordable. Hong Kong's social housing programs have provided nearly 50 percent of that city's total units, and Singapore's public sector has built 80 percent of its housing. Singapore's public housing program, often considered the best in the world, involves careful design to create dense, well-balanced neighborhoods with good public spaces, a mix of building heights, and a mix of residents across classes.[26] The government gave itself authority to buy private land cheaply and today owns 90 percent of the country. Contractors who construct good-quality buildings are rewarded with bidding preferences on future projects.

China's public housing has proceeded in waves influenced by changes in national politics. The public sector built housing in mid- to high-rise buildings between the 1950s and the 1980s, when no private housing development industry existed. Later the government sold public units to their residents, and private developers took over the construction role on land that still officially belongs to the state. The public sector began building housing more actively again after 2007, often capping prices at 70–75 percent of market rates.[27]

Cooperative housing—typically built by nonprofit organizations or the public sector—has been a main way to ensure inclusion of all income groups within urban neighborhoods. In co-op housing, owners or renters are members of the cooperative organization managing the housing and often benefit from low housing prices under a limited equity model in which prices are held below market so that when they sell, owners do not receive a windfall. (New York City has many co-op buildings with apartments, but they tend to be upscale and not designed to promote affordability.) In Sweden, housing cooperatives were started either by citizens wishing to live in such buildings or by unions interested in promoting construction as well as providing good housing for their members. Finland has a somewhat different model; one-third of all housing is owned by nonprofit housing companies in which homeowners hold shares. The national government there also historically made low-interest loans available to developers so as to keep both rental and ownership housing affordable.[28]

In the US, limited-equity cooperatives known as community land trusts are an increasingly popular model. Nonprofit boards composed of community members sell units with deed restrictions so that prices remain below market. Appreciation of home value is limited to a formula that usually approximates inflation, giving residents the advantages of homeownership (greater control over the dwelling, security of tenure, ability to build up financial equity, and ability to benefit from tax deductions for home mortgage interest) but not allowing them the windfall profits from real estate found in cities where housing prices are rapidly escalating. This model could be expanded so that a significant percentage of the housing stock leaves the speculative market. Cities could encourage this by taxing limited-equity properties at a lower rate. The Dudley Street Neighborhood Initiative (DSNI) in Boston is an example of an innovative use of a community land trust to promote affordable housing, develop community capacity, build community wealth, and create green space and community amenities.[29]

Expand Rent Controls, Subsidies, and Tenant Protections

A time-honored if controversial strategy to create affordable housing is for the public sector to regulate rents that private landlords can charge. Typically a municipal government establishes a rent control board that approves annual rent increases of approximately the rate of inflation, with provisions for adjustment if major improvements are made to the unit. Rent control is usually coupled with vacancy control, which means that rents cannot be raised when one tenant leaves and another moves in.

New York City's rent control policies are particularly well known. After the Second World War enormous demand from returning soldiers led to a housing shortage. The city then capped rents for all buildings built before 1947, some two million units, and allowed families to pass the low rents along to subsequent generations. Given the rapid growth of that city's real estate prices, several generations later tenants who kept units in the family were paying a small fraction of market value. Various loopholes allowed landlords to withdraw many units from rent control, meaning that by 2017 only twenty-two thousand units were controlled in this way.[30] A second system, rent stabilization, provided similar controls to certain other older buildings. But newer units rent at market rates, leading to great price disparities. In 2019 statewide legislation gave communities more power to control rents, which may reduce these inequities in the future. Other countries that have strictly capped rents have had similar experiences. Rents have been frozen in Egypt repeatedly over the past one hundred years, leading to a deterioration of building units and decreasing resident mobility, since it is impossible to find any unit as cheap as one that has been in the family for many years.[31]

Landlords and real estate interests argue that controlling rents decreases motivation for anyone to build rental housing in the first place and for landlords to maintain apartments adequately. Yet history shows that rental housing continues to be created in cities with rent control, though perhaps more would have been created otherwise. Rent boards typically require maintenance of rental units, and tenants can file complaints to force repairs.

Cities can potentially craft more nuanced policies in the future to meet the interests of both tenants and landlords. For example, simpler caps on excessive rent increases can help ensure that landlords do not greatly increase rents overnight. California adopted a cap of 5 percent plus inflation for most large apartment buildings in 2019, and Oregon approved a limit of 7 percent plus inflation at around the same time. New York State decided on a broader package of tenant protections that eliminates a number of loopholes that landlords used to raise rents to market values. "Just-cause eviction" ordinances require landlords to have good reason for evicting tenants, such as violation of rental contracts, damaging of property, or drug selling. Other policies ensure that tenants have legal representation in housing court. Policies to limit conversion of rental units to condominiums or Airbnb units can also protect tenants and help reduce the long-term loss of rental units. Matthew Desmond's Eviction Lab has developed a Housing Policy Scorecard to measure how well each state protects tenants' rights.[32] Many progressive states and cities have started to adopt policies to protect tenants' rights and address the national rental housing crisis.

Another tool to assist renters is a straight rent subsidy. This avoids having to regulate the housing market but can be expensive for the public sector. One of the main US housing assistance programs has been Section 8 vouchers, which compensate landlords for rent so that qualifying households have to pay no more than one-third of their income for rent. However, Congress underfunds the program so that only 2.2 million out of some 7.3 million potentially qualifying households in the country receive these vouchers, and waiting lists for Section 8 assistance are long.[33] One relatively simple step to reduce poverty in the US would be to fully fund this program. This would raise costs from about $17.5 billion in 2016 to about $58 billion. Although large, this amount is far smaller than the $60 billion the US spent that year on the mortgage interest tax deduction for homeowners, some 80 percent of which went to households in the top 20 percent of income.[34] In the name of supporting homeownership, US housing policy effectively subsidizes the rich rather than the poor.

Adopt Inclusionary Zoning

Since private developers often aim at upper-income segments of the market, local governments can require them to make a certain percentage of units within each project permanently affordable. Many jurisdictions worldwide have adopted this strategy, often called inclusionary zoning. In Britain, local governments gained the ability to require inclusion of affordable units within development projects under the 1990 Town and Country Planning Act, and by the mid-2000s virtually all large projects were required to create such units.[35] In the United States, some 886 jurisdictions in twenty-five states and the District of Columbia had by the late 2010s adopted inclusionary housing ordinances requiring builders of housing developments above a certain size (typically ten units) to provide a certain percentage of affordable units (typically 10 to 20 percent).[36] Affordability in this case generally means that households making 70 percent of AMI will pay no more than one-third of their income for housing.

The inclusionary housing approach, however, provides only a few lower-priced homes in any given neighborhood. Those are typically inhabited by middle- or working-class households, not extremely low-income households. Developers are also frequently allowed to avoid creating such homes by paying an in-lieu fee to the local government. However, in many cases the local government may then hold such funds for years, meaning that their value diminishes from inflation.

Inclusionary housing requirements can be strengthened to avoid these problems, for example, by requiring some units to be affordable to the very poor making under 50 percent of the AMI and/or to the extremely poor making no more than 30 percent. However, developers frequently argue that they cannot afford to create many such units given the high price of urban land and construction. Hence the need for the public sector and nonprofit sector to step into the housing market in a bigger way. In addition to the mechanisms above, many cities own vacant property that they can offer to developers and nonprofits at low or no cost for guaranteed affordable housing production.

Increase Lower-Half Incomes

One of the most important yet overlooked options to reduce housing affordability problems is simply to raise incomes across the lower half of the economic spectrum. The basic problem is not necessarily that housing costs have risen but that incomes haven't kept up, often because most economic gains have gone to the wealthy. In the US, there is also growing recognition that in order to promote racial and social equity, we need to acknowledge that the Black-white wealth gap is the product of decades of historically racist policies: redlining, discriminatory lending practices, segregated planning policies, employment discrimination, and unfair wage practices. In 2019, the Black-white wealth gap was extreme. The Brookings Institute notes, "The *median* white household held $188,200 in wealth—7.8 times that of the typical Black household." This gap was even greater for Black female-headed households.[37]

Raising wages for lower-end workers is one way to address this problem. However, minimum wage and "living wage" ordinances often aren't enough to help households afford housing in many cities. Wages would need to be far higher to meet current rents and for-sale prices, in the neighborhood of $30, $50, or even $70 an hour in some jurisdictions rather than $15 an hour as proposed by many US living wage campaigns. The MIT Living Wage calculator shows how the living wage varies widely depending on geographic location, number of working adults, and children.[38] Trying to hike wages immediately through national, state, or municipal policy might place excessive burdens on businesses with many low-wage workers, such as restaurants, construction firms, and farms. So a phased escalation of low-end wages may need to be paired with much more progressive tax systems (taxing the poor less and the rich more), greatly expanded social housing, tenant protections, and other mechanisms.

STRATEGIES FOR INCLUSION

Inclusion is about more than ensuring that everyone has decent, affordable housing. Other human needs must be met as well, including needs for health care, quality education (including preschool and college), access to healthy food, a healthy environment, retirement and disability support, safety, and meaningful work. Social safety net programs can meet many of these needs, though such programs can be improved and provided more systematically to everyone. Part-time or contract workers as well as the unemployed in the US, for example, typically do not receive disability or health care benefits and receive only minimal retirement benefits through the federal social security program. Undocumented immigrants often receive no support services at all. Access to many of these societal goods is also determined by where you live. Some towns, cities, and states provide far more and better services than others. In the future to achieve equity we will need to ensure that everyone in societies receives these benefits.

United Nations Sustainable Development Goal 11, which relates to cities, emphasizes inclusiveness: "Make cities and human settlements inclusive, safe, resilient, and sustainable." The term implies special efforts to accommodate people who have historically been excluded. As Black Lives Matter has shown us, structural racism works to exclude people through institutions such as the police, the courts, and prison systems. Exclusion is also woven into urban planning, for example into written and unwritten zoning and other rules that have kept immigrants and people of color out of many neighborhoods.[39] Structural sexism, homophobia, and antidisability discrimination exclude women, gays, and differently abled individuals from full participation in society. All of these forms of exclusion must be overcome (table 4).

Change Institutions

Legal, financial, educational, and economic institutions are a particular focus for change, since they most overtly regulate power in societies. African Americans in particular have been unable to live in many parts of US cities because of racist deed restrictions, large-lot single-family-home zoning, and banks' refusal to give them loans for housing within redlined neighborhoods. The Civil Rights Act of 1964 outlawed discrimination on the basis of race, color, religion, sex, or national origin, though it was not until 2020 that the Supreme Court ruled that the bill's language covered LBGTQ rights as well.[40] The Civil Rights Act of 1968 outlawed overt discrimination in housing. However, discrimination of many types still occurs informally, although it is slowly declining over time.[41] Real estate professionals frequently steer people of color toward certain neighborhoods rather than others, and banks have been found to turn down their loan applications at higher rates or to quote them higher interest rates. Bankers may refuse loans to minority-owned businesses at higher rates than to businesses owned by white men. Wells Fargo settled a lawsuit in 2019 alleging that it had systematically denied loans to minority applicants.[42] Local officials in the US routinely make voting harder in some parts of cities than others, for example by opening fewer polling places or making mail-in ballots difficult to obtain. So the public sector and civil society must work harder to prevent informal discrimination from occurring.

Police forces have received special attention recently. Law enforcement staffs typically embody the same prejudices as the general population and so often treat low-income citizens, people of color, LBGTQ communities, and immigrants differently than others, subjecting them to harsher treatment. One famous example occurred in 2009, when eminent Harvard professor Henry Louis Gates Jr., recipient of fifty-three honorary doctorates, arrived home from a trip to China. Gates, a Black man, found the front door to his Cambridge, Massachusetts, home jammed. He tried to force it open but was then arrested by a police officer who assumed that he was breaking into the house, even though Gates explained that it was his own home.

In 2020 the police killing of George Floyd launched a global social movement calling for institutional changes within policing. Methods include strict accountability for officers, body cameras to record interactions, an end to use of choke holds and other dangerous practices, an end to use of military-type equipment, better officer training, revisions to hiring policies, and police review commissions with power to fire officers and change procedures. It will also be necessary to remove some social service functions from police departments and to eliminate practices that target communities of color within cities and lead to poor relations between the police and local residents. In some cities, disbanding police departments entirely and rebuilding public safety agencies from the ground up may be necessary.

Judicial and prison systems also contribute to inequities and exclusion. African Americans are incarcerated at five times the rate of whites and make up 34 percent of the prison population.[43] Together, African Americans and Hispanics make up 56 percent of those behind bars. Ex-prisoners often have difficulty finding employment and housing because of their records. A decarceration movement, with historic roots going back centuries, is seeking to eliminate or greatly reduce prisons in favor of alternatives such as restorative justice, a set of processes through which victim, wrongdoer, and community members work together to understand a crime's impact, causes, and potential restitution measures. Other programs help individuals and communities overcome the trauma of incarceration. In Philadelphia the People's Paper Co-op helps formerly incarcerated women connect with a network of community organizations and use art to tell their stories and visions for a more just society.[44]

A global recognition that racism and violence against people of color are rampant and unacceptable must lead to a rethinking and reimagining of many institutions that reproduce patterns of racism and inequality. Cities and other levels of government can speed up institutional change through better monitoring of practices and enforcement of anti-discriminatory laws. Basic legislative or constitutional protections may still need to be improved. The goal should be to develop societies that actively undo past wrongs and develop cultures of respect for diversity and inclusion.

Enact Affirmative Action and Reparation Policies

Proactive efforts are needed within sustainable cities to counteract the historical exclusion of disadvantaged groups. Many of these initiatives fall under the label of affirmative action, various types of which have been employed worldwide since the 1960s. In South Africa, for example, employers must create action plans ensuring that they meet target percentages of Black, mixed-race, Indian, female, and disabled individuals at all levels of the organization.[45]

The US federal government and many states and cities require that public agencies take proactive steps to recruit qualified minorities, women, and people with disabilities. For example, women-owned or minority-owned firms often get special consideration for government contracts. Assessments of how representative municipal employees are of the community are demonstrating the need for affirmative action. For instance, the Metropolitan Area Planning Council (MAPC) in Boston produced a 2020 report called "Research: The Diversity Deficit: Municipal Employees in Metro Boston."[46] This project found that the Boston region is drastically missing the mark in terms of representation of people of color and younger people in city government. Boston's former mayor, Marty Walsh, made diversifying the city's public sector a top priority of his administration.[47]

In the late twentieth century affirmative action came under assault from US conservatives, who argued that it discriminated against white people or others from privileged groups. However, the public has increasingly realized that proactive steps are essential to help disadvantaged groups recover from past discrimination and structural barriers within society that have shaped their access to opportunities. New forms of affirmative action may be needed. Some may be formal, such as requirements that any legislature or corporate board be at least 40 percent female and/or racially representative of the community. Others may be less formal. Anyone setting up a committee or panel, for example, can make sure that its membership is diverse. Just look around and ask, "Who is not at the table? Who needs to be here?"

Reparations—lump-sum payments or new funded initiatives to atone for past injustices—represent an even stronger form of action. For example, Angela Glover Blackwell and Michael McAfee argue that banks should be required to provide reparations to African Americans who have been systematically denied access to capital.[48] Others, including The New School Professor Darrick Hamilton, have proposed a government-funded "baby bonds" program to reduce the Black-white wealth gap by establishing an education fund for each Black child. For about $80 billion annually the country could establish savings accounts for infants that would accrue to about $50,000 for low-income children and $200 for wealthy children, significantly narrowing the wealth gap.[49] Other creative reparation schemes could be imagined as well.

Treat Migrants Kindly

Of course, when we think about equity and social, racial, and climate justice, we need to think beyond the nation-state. We are all global citizens. While completely open borders may not be realistic or desirable for most countries, cities can be generous toward those who seek asylum, want to emigrate, or are brought into countries against their will. No more throwing refugees in prison cells or separating parents and children from one another, as done under the Trump administration. National governments can ensure that migrants have access to employment, permanent residency, and pathways to citizenship. Cities can provide them with social services, health care, language classes, shelter, and other necessities of life. In addition to being the right thing to do, providing these benefits will provide numerous co-benefits and cost savings.

Indeed, international human rights law guarantees migrants rights to life, health, and services equivalent to those received by nationals. It also mandates freedom from discrimination, protection against arbitrary arrest and detention, protection against torture or inhuman treatment, and protection against being returned to the country of origin if this would threaten life or freedom.[50] Sweden has been an international example of best practice in refugee treatment, quickly providing permanent residency. It has no mandatory detention of asylum seekers and in fact has encouraged refugees from wars in the Balkans and Syria to emigrate.[51] While the US government has failed miserably in recent years to treat migrants humanely, cities have often stepped up instead. During the Trump administration, Chicago, Portland, Philadelphia, San Francisco, and other cities formed a "Sanctuary City" network to welcome immigrants and refused to cooperate with federal authorities who wished to deport them. An international "Welcoming Cities" network also seeks to develop programs to welcome migrants.[52]

STRATEGIES FOR PROMOTING EQUITY

In 2019 the American Planning Association ratified the *Planning for Equity Policy Guide*. This highlights particular challenges such as gentrification (neighborhood change resulting when new, higher-income residents arrive), environmental justice (disproportionate environmental or health impacts on low-income residents, communities of color, tribes, or immigrants), and community engagement and empowerment.[53] The guide encourages urban planners to take a comprehensive approach to equity problems, listen to local community members, and encourage them to engage in collaborative problem-solving.

These are excellent ways to improve daily planning practice. However, bringing about sustainable cities will require much more fundamental changes as well (table 4).

Adopt More Progressive Tax Policies

Improving the balance of wealth within virtually every society is essential. The most direct way to do this is to change the tax code, which can be done at local and regional as well as higher levels of government. Regional tax-base sharing, for example, is an underused option to promote more equitable access to amenities such as schools and parks across metropolitan regions. The Minneapolis–St. Paul region is the only US urban area to do this so far, with local jurisdictions contributing 40 percent of tax revenue from industrial and commercial development since 1971 to a regional pool.[54]

Progressive rate structures are needed in which marginal tax rates (the rate that people pay on income above a certain level) rise steeply with income and at the top approach 100 percent. Income taxes are applied primarily at state and national levels, although seventeen US states allow cities to levy them as well. Detroit, for example, imposes a 1.2 percent income tax on nonresidents employed in the city.[55] However, progressive rate structures could be applied to property taxes as well, which are what local US governments depend most heavily on. Such a step has precedents in Europe and has been proposed in British Columbia in the form of a rising surcharge at increments of 0.5 percent of assessed value for homes worth more than $1 million, $1.5 million, $2 million, and $3 million.[56]

Many countries had highly progressive tax rates in the 1950s and '60s. The top US income tax rate was 91 percent or above between 1951 and 1963, during the Eisenhower and Kennedy administrations.[57] Yet more than fifty years of neoliberal government reduced this rate to 37 percent by 2020. The top UK rate peaked at 99 percent in 1945 and was 83 percent as late as 1979.[58] Even in the middle of the relatively low-tax era of the 2010s, Sweden had a top marginal rate of 57 percent, Denmark a rate of 55 percent, the Netherlands and Finland rates of 52 percent, Japan a rate of 51 percent, and Israel a rate of 50 percent.[59] Meanwhile, households making less than, say, $50,000 could be exempted from taxation. Making tax schedules more highly progressive would generate large sums for public services, much of which could then be rebated to local government on the basis of population and need, thus helping equalize resources across jurisdictions.

However, since the rich in capitalist countries earn most of their money through capital gains (returns on investments) rather than income (salaries), raising capital gains taxes to similar levels is also necessary. Reducing capital gains taxes has been a main goal of neoliberal governments and serves mainly to enrich the wealthy. Better yet, a wealth tax could proactively start reducing the great fortunes that currently exist. US senator Elizabeth Warren introduced legislation in 2019 to create a 2 percent tax on household wealth above $50 million, with an additional 1 percent on fortunes above $1 billion.[60] Senator Bernie Sanders and others have made similar proposals.

To reduce accumulation of vast family fortunes, the public sector will also need to make inheritance taxes much higher. This form of taxation has been continually weakened in the US to the point where the first $11 million of an individual's estate is passed on tax-free. Structural racism, including historic and present difficulties in buying property in many locales, has prevented African Americans from accumulating family wealth in this way. A 2016 Brookings study found that the average white household in the US had almost ten times more wealth than the average Black household.[61]

To improve the economic situation of those at the bottom end of the spectrum, cities can set minimum wages substantially above state or federal levels. They can adopt their own, highly progressive income taxes, rather than relying on regressive sales taxes for much of their revenue. State or federal governments can require employers to limit maximum salaries to, say, ten times their minimum salary. (Currently the average ratio in the US is around 70 to 1, with some CEOs making over three hundred times as much as the lowest-paid worker.)[62] Given that executives would want to keep their own pay high, this might lead to rapid increases in worker pay.

To truly eliminate poverty, guaranteeing a universal basic income may be needed if other social safety net programs aren't improved. This approach has been widely considered for several hundred years and gained some traction among candidates during the US Democratic primaries in 2019–20. It is called by different names worldwide, such as "guaranteed minimum income," "citizens' income," "living stipend," or "negative income tax." Each resident of a given city, state, or nation would be assured of sufficient financial resources to be able to afford the basics of life, including food, housing, and health care. Homelessness, at least due to lack of financial resources, would become a thing of the past.

A number of short, small-scale experiments have been done with universal basic income. In the US, one New Jersey trial enrolled 1,357 families between 1968 and 1972, and another in Gary, Indiana, enlisted 1,780 households between 1971 and 1974. Although economists have worried that guaranteed income would reduce people's incentive to work, these experiments found that men reduced their work by the equivalent of only two weeks a year, and women by three[63]—to about the level of a full-time job with sensible vacation benefits such as enjoyed by many Europeans already. A similar Canadian experiment in Manitoba showed a negligible effect on work hours, with men reducing hours 1 percent, married women 3 percent, and unmarried women 5 percent.[64]

Some critics argue that a universal basic income sufficient to end poverty in the US would be enormously expensive.[65] But once again this is a question of social priorities. Providing a supplemental income of, say, $20,000 annually to the twenty-five million poorest US households (containing fifty to seventy-five million residents) would cost $500 billion annually—an enormous amount of money, certainly, but less than 40 percent of the country's $1.25 trillion total annual spending on wars, the military, intelligence, and homeland security.[66] To end homelessness and extreme poverty, such an investment would be well worth it.

Ensure Spatial Equity

The rapid spatial fragmentation of metropolitan regions worldwide that has taken place over the past century works against equitable and inclusive societies. With rapid suburbanization, people of different backgrounds have often moved to communities miles away from each other, with vastly different access to amenities and services, particularly schools. Some affluent households choose gated enclaves in which they rarely have to encounter others different from themselves. Since property tax base, school quality, and local parks and services vary by jurisdiction, this means a highly inequitable access to opportunity and negative impacts on society generally.

Metropolitan regions are not going to change their physical form overnight. However, good, equitable, and coordinated planning can increase opportunities for low-income individuals and people of color to live in any community and feel comfortable everywhere. Cities can change zoning codes so as to require that developers create a range of housing types and sizes within any new neighborhood to accommodate residents from different income groups. They can prohibit gated and large-lot communities. They can end single-family zoning within existing neighborhoods, as Minneapolis has done, to promote diversification of housing types and resident incomes. Local governments can rezone old shopping malls and industrial sites for new, diverse housing and can monitor real estate and lending institutions to catch any remaining discriminatory practices.

The design of public spaces can seek to honor all local cultures, creating an atmosphere in which marginalized residents feel more comfortable. Local artists of color can be hired to design these spaces. Murals can portray diverse local cultures and histories. Programming of public spaces can emphasize culturally diverse performances.

Protect Vulnerable Residents from Displacement

Sustainable cities must address the problem of gentrification. In almost any city that is successfully revitalizing older neighborhoods, existing low-income residents and/or people of color are displaced. The culture and political control of neighborhoods may also shift toward wealthier white newcomers.

Addition of higher-income residents is not bad for neighborhoods per se and can be highly positive if it brings more jobs, more amenities, safer streets, and stronger connections between people.[67] The challenge is to eliminate displacement of longtime residents through some of the mechanisms previously discussed.[68] Protecting existing renters by capping rent increases and requiring just-cause reasons for eviction is a starting point. Helping low-income homeowners get access to home equity lines of credit may be important for upgrades and maintenance. Capping property tax increases for low-income homeowners may enable them to stay in their homes. Commercial rent control or other support for existing small businesses, particularly Black- and Brown-owned businesses, may be necessary as well. The effectiveness of such protections often depends on initiating them before development pressures get too severe—that is, planning.

Cities can proactively manage neighborhood change by creating a Neighborhood Plan or Specific Area Plan and adopting zoning code changes if necessary. Such processes can establish a vision of incremental, culturally compatible change. They may be combined with a form-based code that establishes parameters for new buildings, for example, requiring them to have footprints of less than a certain size so as to fit into the existing urban fabric. The scale and pace of new development may need to be limited so that change happens in an incremental fashion, rather than as a tsunami of new building. Measures to protect historic buildings may be needed.

A plan for a neighborhood at risk of gentrification should include strategies to expand jobs for existing residents. New businesses can be required to hire a certain percentage of local residents as well as women and people of color. Training programs can be established for local high school graduates. Municipal contracts can be given to businesses owned within the community rather than to firms from farther away. Through such policies, neighborhood change can provide existing residents with enhanced employment opportunities.

Retrain Ourselves

Policy strategies and planning to promote equity are not the whole picture. Personal change within each of us is required as well. In our workplaces and daily behavior, we can identify small ways to help create more diverse, inclusive, and equitable communities. We can ensure that every event and organization we're associated with is open and welcoming to diverse others. We can call out microaggressions (for example, "You're smart for a woman," or "You don't speak the way I thought you would," or "Where are you really from?") We can suggest more inclusive processes. We can build friendships and networks with diverse others.

Educational systems at all levels can emphasize the study of historic injustices and the efforts to address them. They can help students understand interpersonal dynamics and employ respectful communication. Workplaces can conduct diversity and antiracism training. They can also revise policies so as to better support female employees and families, for example by providing access to childcare and eldercare and paid time off. The language and images in publications and on websites can be improved, as well as the diversity of speakers at events. Hiring, retention, and promotion processes can be examined to see if they subtly or not-so-subtly discriminate against certain types of applicants, preventing them from full participation in the workforce.

CONCLUSION

Making cities more affordable, inclusive, and equitable requires a complete rethinking of urban systems and how they are nested within larger economic and political systems. Exact strategies will vary from place to place. But the starting point is to acknowledge the ways that many of our existing systems perpetuate inequity and suffering of various types. Everyone in society can be oppressed by these systems. For example, although non-Hispanic white men are the most privileged group in American society, they accounted for 81 percent of US suicides in 2017, a number that had risen 40 percent since 1999.[69] Although they have plenty of privilege (particularly those who are not low income), white men are clearly still suffering. A more just and equitable society will potentially improve their lives as well.

It is time to imagine cities in which social and racial and gender equity, tolerance, inclusion, and compassion for one another are at the center of public life, in which racism, discrimination, and intolerance are relics of the past, and in which we change not just our language but our behavior and institutions. Black Lives Matter and other recent social justice movements challenge us to make these changes. It is time to step up and organize our peers, colleagues, coworkers, families, and neighbors to do so.

5

How Can We Reduce Spatial Inequality?

While this book is about reimagining sustainable cities, we pause here to connect sustainable cities with the larger national and international context in terms of spatial inequality. We live in a world that is deeply interconnected. If we want sustainable cities, we need to work on reducing spatial disparities between cities and rural areas, and between different regions worldwide. Linkages between communities need to be recognized, and resources shared and equalized. Situations must be ended in which some regions exploit others by giving them the unwanted by-products of production, such as pollution, waste, and labor exploitation, while simultaneously moving resources and profits from poor regions to rich ones.

In and around the towns of eastern Kentucky, where Stephen Wheeler's ancestral family is from, people of English and Scottish descent lived for many generations as self-sufficient farming families. That way of life changed in the second half of the twentieth century. Better roads, electricity, and telecommunications connected Appalachia with the rest of the world. Urban job opportunities lured away the young. Farming families became part of the cash economy and acquired new desires for processed foods, appliances, motor vehicles, and personal accessories. But hill farms didn't generate enough cash to buy such things, especially with rising federal subsidies for agribusiness in other parts of the country. So the people of eastern Kentucky became designated as poor and came to see themselves that way.

Environmental problems grew as well. Giant bulldozers scraped away hilltops and extracted coal, adding this region to the long list of others worldwide suffering from the "resource curse." Runoff from coal mining poisoned wells and polluted waterways. Coal jobs left as quickly as they had come, leaving many even poorer.

A new, more globalized retail economy brought first Kmart and then Walmart, putting family-owned stores out of business. Fast-food outlets proliferated. But the new service economy jobs didn't pay much. To make better money some people began growing marijuana in hard-to-reach locations in the hills. Drug use, alcoholism, and obesity spread. Fundamentalist religion gained adherents and combined with Fox News (starting in the 1990s) to promote reactionary political values. A region that had been Democratic until the late twentieth century now helped elect US Senate majority leader Mitch McConnell (R-KY). McConnell in turn played one of the largest roles in thwarting progressive legislation from Barack Obama's administration, supporting Donald Trump's presidency and fueling the rise of populism in the US.

If this tale of decline were one isolated example, it might not matter much. But spatial inequality persists and spreads worldwide. Some left-behind communities are rural. Others are urban. Entire countries are stuck in poverty due to the legacy of military or economic colonization.

Spatial inequality is a core challenge to the development of more sustainable cities. Every community needs to be able to thrive, not just certain favored ones within a highly unequal global system. Instead of engaging in a zero-sum approach to development, with winners and losers, communities need to support one another so that all improve their quality of life and sustainability.

The so-called winners of today's global economic competition have their own problems. At the other end of the spectrum from Appalachia is Silicon Valley. This forty-mile corridor in the San Francisco Bay Area is an economic dynamo envied the world over. Covered by orchards and agricultural fields in the 1950s, this beautiful area was known as "Valley of Heart's Desire." Now no orchards remain, and the region is a congested sprawl of poorly connected office parks, subdivisions, malls, and commercial strips. Incomes are high, but the price of a home is nearly five times that in the US as a whole.[1] Many residents cannot afford housing near their jobs and so endure lengthy commutes or are housing insecure. Social inequality, traffic congestion, air pollution, and greenhouse gas emissions expanded greatly during the past fifty years, reducing the quality of life in the region and contributing to global warming.

The Silicon Valley ethic of "move fast and break things" has created dynamic companies, unprecedented technology, and great wealth for a few. But the new gig economy pioneered there often operates at the expense of workers and the environment. It often produces an enormous concentration of wealth that comes from the exploitation of others. One study found that one-fifth of San Francisco Uber and Lyft drivers earned virtually nothing when their full expenses, including things such as health insurance, were accounted for.[2] The tech industry has also been heavily criticized for sexual harassment during the MeToo movement and racism during the Black Lives Matter movement. The combination of individualism, predatory capitalism, toxic masculinity, and lack of concern for the common good that Silicon Valley represents works strongly against a sustainable and equitable future.

Similar problems of unequal development exist in other successful urban areas worldwide, including Shanghai, Beijing, Tokyo, Bangalore, Singapore, Toronto, London, Amsterdam, Paris, and Tel Aviv. Though among the world's economic success stories, on many dimensions of sustainability they are failures. The growing core-periphery disparities that produce left-behind communities and "sacrifice zones" on the one hand and wealthy but unsustainable and highly unequal job centers on the other are at the heart of recent global development patterns.

Let us imagine instead a world where we are *not* content with the concentration of wealth and opportunity in a small number of global cities; where all communities have affordable housing and provide a decent quality of life; where cities meet the needs of people locally and regionally but do not drain wealth from other parts of the world; where no areas are left behind in the transition to a green economy, their populations increasingly alienated, despairing, and vulnerable to unscrupulous politicians and warlords; and where social dimensions of sustainability are well served everywhere.

SOURCES OF THE PROBLEM

Today's spatial inequity problems have long historical roots, illuminated by literature in fields such as economic geography, sociology, and environmental history. One starting point is physical geography. Some parts of the world have more fertile soils than others, more abundant mineral resources, more useful plant, animal, and fish species, and/or more benign topography and climate. Other places have been strategically well located to serve as trading centers and market towns or have been easy to defend against attack. Such communities have been able to accumulate modest amounts of wealth and power. The "chessboard" of geographical wealth is constantly shifting and with global warming is likely to shift in even greater ways in the future.

However, in other cases spatial inequities have resulted from military, religious, cultural, political, and/or economic systems that further centralize power and wealth. Typically these have drained resources from the periphery to the core of empires. Many parts of the world still suffer the legacy of colonization. Local traditions and cultures were disrupted, peoples were exploited, racism was institutionalized, ecosystems were harmed, and corrupt, colonizer-friendly governments were installed following independence. The damage has been so profound and long-lasting in many places that reparations may be appropriate. The need for climate justice may likewise call for reparations and repayments.

Twentieth-century economic development philosophies exacerbated spatial inequality on the assumption that economic globalization was to everyone's long-term benefit. Various versions of "growth pole" theory, originating in the 1950s, sought to focus business development in particular geographical locales within countries on the assumption that this would leverage economic development in other parts. Such wider-scale progress was rare; growth poles instead often channeled resources to local elites, created isolated business enclaves, and harmed the environment.[3]

The municipal economic development practice of chasing branches of multinational corporations has likewise undermined prospects for a more stable long-term economic base in cities worldwide.[4] This "race to the bottom" competition leads suburbs to compete to host the newest shopping mall, central cities to compete for corporate headquarters, and states or countries to lower their environmental and labor standards to attract multinational corporations. However, the resulting businesses often don't provide the expected number of jobs, pay the decent wages promised, or stay more than a few years.[5] As Margaret Dewar has pointed out in her well-titled article "Why State and Local Economic Development Programs Cause So Little Economic Development," politicians have an incentive in the short term to appear to be generating jobs by attracting well-known companies but little incentive to take into account long-term economic or environmental sustainability.[6] A recent example of the extreme lengths that municipalities will go to in order to attract development can be seen in the global competition for the second Amazon headquarters.[7]

The Bretton Woods framework of post–World War II development assistance only deepened global spatial disparities, creating what economist Andres Gunder Frank termed "the development of underdevelopment."[8] Agencies such as the World Bank and the International Monetary Fund loaned funds to developing countries for megaprojects that created wealth for elites but left others poor and displaced, while countries accumulated enormous debt to lenders in the Global North. National governments focused on what sustainability-oriented NGOs refer to as "extreme infrastructure."[9] These dams, power plants, industrial zones, and large-scale agricultural projects sought to jump-start an export-oriented form of economic development that was often environmentally harmful and funneled capital created by Third World labor and resources into First World bank accounts.[10]

Yet another source of disparities has been the structural adjustment policies that neoliberal governments in wealthy nations insisted upon as a condition for international assistance during the past forty years. These require developing countries to take actions such as cutting social programs, privatizing public assets such as utilities and railroads, reducing barriers to foreign investment, and lowering taxes on the wealthy. The effect has been to make life harder for the poor while enriching elites and international corporations. It is increasingly clear that structural adjustment policies need to be discontinued and policies that promote spatial equity put in their place.

Finally, the offshoring of manufacturing from wealthy nations to low-cost and less regulated parts of the globe during the past half century has had complex effects on spatial disparities. It has impoverished the US Rust Belt as well as the British Midlands, leading to the growth of right-wing populism in both places. Meanwhile, it has helped fuel the rise of megacities and megaregions in the developing world, leading to massive internal migration and expanding economic disparities between those urban areas and the countryside. Undoubtedly, these global economic shifts have improved quality of life for many. But they have harmed others, disrupted societies, contributed to the climate crisis, and widened the gulf between rich and poor communities (figure 7).

Although spatial disparities are still expanding in many places, there is hope for the rebirth of left-behind cities and regions. Manchester, UK, the first industrial powerhouse in Europe, lost much of its manufacturing in the middle of the twentieth century but has since rebuilt itself by focusing on culture, education, physical regeneration, and its geographical role as a transportation center. The US steel capital of Pittsburgh,

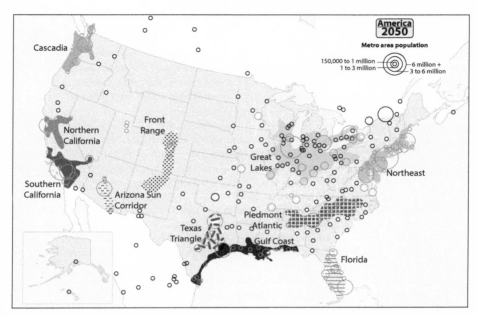

FIGURE 7. The benefits of global economic development flow increasingly to certain constellations of urban regions worldwide that have been called "megaregions." Meanwhile, other areas are left behind with growing poverty. (Regional Plan Association, "American 2050: A Prospectus," New York, September 2006, 10, https://s3.us-east-1.amazonaws.com/rpa-org/pdfs/2050-Prospectus.pdf.)

Pennsylvania, after losing 350,000 industrial jobs in the 1980s, reinvented itself as a center of renewable energy, health care, and education.[11] Even the long-declining hulk of Detroit, one of the most hollowed-out American cities, is showing signs of a turnaround. Examples such as these indicate the possibility for left-behind places to rebound. But all of these cities had assets to start with, including a strong identity and an active elite that led revitalization efforts. Other communities and regions don't have such advantages. And the pervasive problems associated with spatial inequality affect wealthy as well as declining places, necessitating holistic and imaginative solutions at higher levels of governance.

STRATEGIES TO REDUCE SPATIAL DISPARITIES

We might easily throw up our hands when contemplating entrenched spatial inequities. How can these ever change? Yet creative, synergistic actions can be undertaken by many different players (table 5). Collectively we need to reimagine spatial geographies of development that

TABLE 5 STRATEGIES TO REDUCE SPATIAL INEQUALITY

Strategy	Description
Ensure basic rights everywhere.	Work at higher levels of government to ensure basic human rights within all communities and regions. Strengthen international justice frameworks, national constitutions, national/state legal protections, and local enforcement.
Equalize funding and resources.	Have higher levels of government ensure a baseline level of resources per capita in addition to need-based funding. Promote tax-base sharing between cities and towns within urban regions. Evaluate the spatial equity impacts of major public investments. Consider a global wealth tax, distributed on the basis of population.
Target additional assistance to those who most need it.	Provide resources (financial, institutional, technical, human) from higher levels of government and civil society to disadvantaged communities. Strengthen NGOs that assist disadvantaged communities. Avoid creating marginalized communities in the first place.
Ramp up reparations.	Provide assistance that specifically attempts to correct for historic injustices, for example through needs-based investments in schools, health care, social services, and housing. Reform institutions such as police and prison systems that have historically oppressed particular groups. Consider providing individuals with subsidized home mortgages, money that could be used to support their education, or cash payments.
Bring about structural change.	Rethink underlying systems that create spatial disparities, especially neoliberal capitalism. Restrict flows of capital from poor nations or regions to rich ones. Require businesses to reinvest profits in local communities. Ensure that companies mitigate negative environmental and social impacts of their operations. Promote values of mutual assistance, solidarity, and cooperation.

build networks of support among sustainable cities and regions. The transition to a green economy, talked about in Green New Deal proposals, provides new opportunities for spatial equity. Other opportunities for reimagining spatial equity come from the growing ability of people to work remotely and live in lower-cost, more climate-ready regions.

Many interventions will depend on coordinated, multiscalar action by higher levels of government or international cooperation. Local political systems often lack the resources, authority, political will, or breadth of vision to pull themselves out of difficult situations or assist their neighbors. So states, regions, national governments, or international agencies must step in to provide the needed funds, technical assistance, or institutional reforms. Old industrial cities in the US, for example, have benefited from the federal Superfund program, which has cleaned up more than 424 sites with serious toxic waste contamination since 1980.[12] Other federal programs such as Section 8 housing vouchers (subsidizing rent for low-income people), the Supplemental Nutritional Assistance Program (SNAP, subsidizing food), and the Earned Income Tax Credit (which provides additional income to low-income workers) benefit distressed communities more than others.

Other interventions are provided by civil society, including organizations such as the United Nations, Red Cross, Oxfam, and the International Rescue Committee. The latter organization, for example, works in communities in both poor and wealthy nations to help people harmed by warfare, natural disasters, or displacement. The fact that it is headed by David Miliband, a former foreign secretary and environment secretary for the United Kingdom, shows how such nongovernmental organizations have gained a stature equivalent to that of public sector agencies. Indeed, many NGOs now do what in the era of neoliberalization the public sector has been unable to in terms of assisting communities and regions hardest hit by the side effects of economic globalization. But the future problems we face will need government investment: we cannot rely on NGOs to fill the void left by neoliberal governments.

Still other actions take place at a personal level. For example, many residents, cafes, and co-ops in North America and Europe now purchase fair trade coffee produced by small-scale farmers in struggling regions such as Central America and East Africa. NGO intermediaries guarantee farmers decent prices for their coffee; other nonprofits such as Fairtrade International certify these transactions.[13] This new economic model is based on the concept of a "social premium," an extra amount that buyers pay to ensure that their purchases promote environmental and socially responsible societies. Other examples include consumer purchases of sustainably harvested lumber, recycled products, and renewably generated electricity. These approaches must be scaled up so that they push markets to provide affordable and sustainable products.

Ensure Basic Rights Everywhere

A starting point for reducing spatial inequality is to ensure that people everywhere have the same basic human rights. Humanity has acknowledged this need for more than seventy years, since United Nations approval of the Universal Declaration of Human Rights in 1948. This document in turn built upon previous agreements such as the Geneva Conventions regarding the treatment of prisoners of war and civilians in wartime, the first of which dates to 1864. And those in turn built upon long-standing ethical beliefs in "treating others as yourself" within many cultures and religions.

Human rights are first and foremost guaranteed by national constitutions. As these documents have been created or revised over the past fifty years, they have increasingly provided a baseline of rights for individuals in countries worldwide. For example, 76 percent of national governing frameworks now explicitly protect against discrimination on the basis of race and ethnicity.[14] Eighty-six percent now specifically promise equal rights to women.[15] Social safety net policies can also be considered human rights and another important step toward reducing disparities. Article 22 of the UN Universal Declaration on Human Rights states that "everyone, as a member of society, has the right to social security," while Article 25 states that "everyone has the right to a standard of living adequate for the health and well-being of himself and of his family, including food, clothing, housing and medical care and necessary social services, and the right to security in the event of unemployment, sickness, disability, widowhood, old age or other lack of livelihood in circumstances beyond his control."[16]

Articulating these rights helps everyone and is an enormous step forward historically but makes the greatest difference to disadvantaged communities. Such principles are the foundation for more equally sustainable cities and regions in the future. However, without strong action to translate them into practice they may be meaningless. Entire economic, social, and political systems are designed to keep certain groups in power and exclude others. Social movements have had to arise over and over again to demand better treatment for all. The Black Lives Matter movement, for example, is just the latest installment of more than three hundred years of efforts to bring African Americans to full equality. Other movements for women's rights, gay and transgender rights, workers' rights, and the rights of ethnic and religious minorities have likewise sought to bring promises of human rights into reality.

Another frequent strategy for implementation has been to ask higher levels of government to intervene. US civil rights leaders were forced to do this in 1957, when they asked the federal government to send troops to enforce the *Brown v. Board of Education* Supreme Court decision mandating desegregation of southern schools. Although the State of

Arkansas was staunchly opposed to desegregation, President Dwight Eisenhower sent units of the National Guard to escort nine brave Black students into Little Rock High School. In many less dramatic ways, national legislation, regulation, and enforcement bring human rights into practice within local communities otherwise resistant to them.

Yet there is no world government with power to step in and guarantee human rights in most cases where entire countries fail to provide them. The United Nations' ability to do so is notoriously weak, and UN peacekeepers typically intervene only to keep the peace after a military conflict within a country.

Still, nations, NGOs, and international agencies can help countries do a better job of bringing human rights and environmental protections into reality through means such as diplomatic encouragement, citizen pressure, sanctions, financial incentives, international agreements and protocols, and boycotts. Recent international campaigns led primarily by NGOs have sought to put pressure on China for its oppression of Tibetan and Uyghur minorities, on Israel for its occupation of Arab lands and oppression of Palestinians, and on Saudi Arabia for its oppression of women and minorities as well as its military violence against Yemen.

Perhaps the most famous success of international pressure, involving state and UN actors as well as NGOs, was the ending of apartheid in South Africa in 1991. Although many internal factors also helped bring down this system of strict racial segregation, including the dedicated struggle of activists such as Nelson Mandela, international pressure played a main role. United Nations resolutions, cultural and economic boycotts, worldwide student protests, direct financial assistance to the opposition African National Congress by the Nordic countries, and moral opprobrium expressed through the increasingly global media of the time all helped end the subjugation of some thirty-five million Black and mixed-race South Africans by about four million whites.[17]

In the future, international bodies such as the United Nations Commission on Human Rights could play a much larger role in investigating and publicizing rights abuses around the world and adjudicating and incentivizing climate justice. Nations could do a better job of initiating and coordinating sanctions. The growing body of NGOs worldwide can help rally pressure and provide direct assistance. Cities and states can even boycott others if necessary (as of this writing, California does not allow its employees to use state money for work travel to a dozen other states such as South Carolina because of their poor human rights records). Businesses can also play a role in bringing pressure to bear, and shareholders, employees, and customers can pressure them to do so. An example of this occurred in 2020 when corporations boycotted Facebook for its tolerance of hate speech.

Equalize Funding and Resources

Official guarantees of rights will not mean much to a young woman in New Orleans, Tegucigalpa, Lagos, or Bangkok if she is impoverished, has no access to good-quality education, is from a minority group within society, has no reproductive rights, or lives in a culture in which women are expected to stay at home or undertake only menial jobs. Her situation will change only if she and others around her have opportunities for a better life and if the public sector has the capacity to provide those and acts with care and compassion. Amartya Sen, the Nobel Prize–winning Indian economist who is one of the leading philosophers of the past hundred years, argues that freedom is meaningful only if people throughout human societies have the capabilities to take advantage of it.[18] We need to reimagine a world that proactively provides this type of access to opportunities.

For members of disadvantaged communities to have these opportunities, higher levels of government will often need to step in with support and funding. Some interventions will need to be regulatory, for example to prohibit discrimination in the workplace or in bank lending. Some will need to be cultural, for example to educate men so as to eliminate domestic abuse and patriarchal restrictions on women's freedom. Some may involve educating people about being better earth stewards or demanding socially responsible products. But many will involve consistent funding. Nations, states, and provinces will in effect need to transfer resources from wealthy communities to poor ones. In a broad sense this process is known as equalization, and it plays out in several different ways.

One form of equalization occurs by default as higher-level governments distribute need-based funding for services or infrastructure to lower levels of government. Disadvantaged communities naturally receive more per capita; those with greater wealth and lower need receive less and on balance are net contributors. For example, Northern Ireland, Wales, and Scotland—all economically depressed regions of Great Britain—each receive at least £2,800 per capita more from the central British government than they contribute in taxes.[19] In contrast, relatively affluent London contributes a net of £3,070 per capita. Similar transfers occur within most other countries and at the state or provincial level as well. However, they are often not nearly enough to reduce disparities in our highly unequal world.

Another form of equalization occurs when a national government makes cash payments to underresourced parts of a country so that residents of those communities do not face higher tax burdens for a given level of public services. Canada, Australia, Germany, and Switzerland all have long-standing programs of this sort, known as fiscal equalization. Under the Australian system, for example, the less advantaged state of Tasmania receives payments making its budget 1.8 times as large as it would be otherwise.[20] Wealthier states receive no such funding. Switzerland has had programs to level out the tax base between its cantons since 1938. Rather than taking equalization monies out of general funds, it requires wealthier cantons to contribute specific amounts into a pool that is then redistributed to the less well-off.[21]

Within the United States, the Community Development Block Grant (CDBG) program, initiated in 1975, has some equalization elements in that this federal program distributes funds to local governments under a formula that takes into account both the locality's share of the population and its level of need. State and local governments have relative freedom to use CDBG monies as they see fit to meet local needs. However, CDGB funding is now less than a quarter of its original size, and the 2020 level of $3.3 billion is very little for a county of 332 million people—barely ten dollars per capita.[22] Such an inconsequential amount is nowhere near enough to equalize financial disparities or address pressing local needs. Raising this federal funding by a factor of 100—still less than a third of the annual US military budget—might begin to meet both objectives. At a level of $1,000 per capita annually, cities in need might then have budgets 50 to 100 percent higher to meet local needs.

At the state level in the US, equalization programs often focus on education. K-12 schools are run by local governments and funded by property taxes, and local fiscal capacity varies widely. One study found that property value per capita for a typical wealthy Massachusetts suburb was 6.2 times that of a large city in the same state.[23] Racial divisions mirror those of class; another study found that white school districts in the US spend $2,226 more per pupil on average than nonwhite school districts and that in Arizona the disparity is almost $11,000 per student.[24] Those disparities matter greatly in terms of opportunities to students in different cities and neighborhoods. As a result of citizen lawsuits, all but five US states have adopted equalization formulas that attempt to reduce per-pupil funding disparities, usually by setting a minimum funding level.[25] However, the amounts of state and federal equalization funding are small in proportion to school budgets (the federal contribution is about 10 percent) and make only a modest difference in reducing disparities. Much greater equalization funding is needed for equitable education.

Sharing tax base across jurisdictions is yet another way to reduce spatial fiscal inequities. As mentioned previously, this has been done to a limited extent in the Minneapolis–St. Paul metropolitan region since 1971. A far more radical approach would be to centralize all tax collection at the highest practical level of government and then redistribute much of it on the basis of population with additional funds on the basis of need. "From each according to his ability and to each according to his need" was the socialist slogan popularized by Marx in 1875.[26] After the extreme inequality of recent decades, this principle's time may finally have come. Such a strategy would greatly reduce disparities between communities, give all individuals more equal opportunities, help address racial wealth gaps, and eliminate the current hodge-podge of tax-collecting entities that exists in most countries across scales of government.

Unfortunately, some higher-level governmental policies worsen spatial disparities rather than improve them. Building a freeway system or high-speed rail network, for example, tends to enrich landowners and businesses near interchanges or stations unless measures are taken to recapture this unearned windfall for the public. US military spending stimulated growth of the country's South and West during the second half of the twentieth century, through contracts allocated disproportionately to California, Arizona, and Texas.[27] These spatial effects were probably not deliberate but were nonetheless profound, helping position those regions at the forefront of new electronics industries. Considering the spatial, intersectional and equity implications of public sector decisions in advance could lead toward fairer and more balanced forms of development. Preparing social impact statements for major programs and policies is one way for cities and societies to do this. Creative planning approaches can be used to capture the benefit of development and redistribute it more equally; however, to do this we need to make addressing spatial inequality a priority.

There is, alas, currently no mechanism for equalization on a global scale. Many wealthy nations contribute modest sums to multilateral or bilateral foreign assistance programs, but these amounts are small compared to the size of their economies. Such donations are also often under the control of the donor nations, come with strings attached such as a requirement that recipients buy donor products, and are inadequate to address vast global disparities in resources. Much broader mechanisms are needed, especially as climate justice needs become more dramatic. The Clean Development Mechanism established under the 1997 Kyoto Protocol was intended as one such international climate equity process. Industrialized nations were allowed credits toward their GHG reduction commitments by funding emission reduction projects in developing nations. However, this market mechanism has never functioned well, and enormous questions exist about whether the funded projects really reduced emissions in the promised ways.[28]

Perhaps someday a global wealth tax, administered by the United Nations or some other democratically run agency, will exist. It could apply progressive tax rates equally to every global citizen and then make equalization payments to nations and their local governments. Such a program would need to be structured carefully to ensure that funds actually benefited communities at a grassroots level. But if done well, the effect would be to redistribute wealth from affluent nations to poor ones and to give residents of disadvantaged communities worldwide far greater opportunities for education, improved health care, better housing, and economic opportunity. Such a process could start on a small scale with the Tobin tax on global financial transactions discussed previously.

Target Additional Assistance to Those Who Most Need It

More specific programs of assistance for disadvantaged communities will be needed as well. The philosophical justification for such assistance has long existed. Most religious traditions have promoted concern for the poor, and "treating others as oneself" has been central to Western ethics as well, expressed for example through Kant's categorical imperative and John Rawls's argument that concern for the least well-off should be the cornerstone of modern ethics.

Targeted assistance for disadvantaged communities and regions is already common worldwide. The US federal government has tried to assist Appalachia through agencies such as the Rural Electrification Administration, the Appalachian Regional Commission, and the Tennessee Valley Authority. The European Union has devoted tens of billions of euros to assisting the south of Italy (the Mezzogiorno) and other poor regions. International development agencies have tried to assist Africa, Latin America, and South Asia ever since the late 1940s. These initiatives have often focused on building infrastructure and have had only limited success. In the case of Appalachia, new dams harmed the ecology of watersheds, and new roads facilitated exploitative industries such as logging and coal mining.

A better approach would be to change the national and global policies that help marginalize regions in the first place and tying them to greening policies. For Appalachia, this might mean reducing subsidies for midwestern and California agriculture.[29] For many developing nations, this might mean steps to prevent multinational corporations from exploiting local communities and ecosystems while exporting capital. In the case of the Mezzogiorno, underdevelopment is likely due to long-term structural conditions including clientelist politics, Mafia corruption, the absence of civic culture, exploitation by the North, a tradition of large landholdings, and globalization undermining the Italian small-firm model.[30] Specific policy changes and enforcement addressing these issues would be more effective than infrastructure funding.

Assisting left-behind cities and regions requires redefining "development." Local governments as well as outside players could focus on improving the education, skills, and self-organization of residents.[31] They could seek to improve institutional and social capacity, accountability, democratic function, local control, and leadership.[32] Following the principles of Asset-Based Community Development (discussed in chapter 3), they could build on existing community assets to support small, local businesses and improve services and quality of life. Large infrastructure projects and massive loans for business recruitment are less important.

NGOs large and small have created a global network of civil society devoted to improving conditions for disadvantaged populations. They also create opportunities for person-to-person development assistance. Since 2004, for example, the Global Brigades organization has brought more than eighty thousand volunteers from developed countries to 492 partner communities in six developing nations to build medical and dental clinics, construct clean drinking water systems, and promote grassroots economic development.[33] Habitat for Humanity, famously championed by former US president Carter, has helped provide housing for more than thirty-five million people in seventy countries since 1976.[34] Many faith-based networks perform a similar function. The Catholic Volunteer Network, for example, links nearly twenty-eight thousand volunteers annually to some two hundred local programs in 109 countries and forty-five states.[35] Citizens of well-off societies also assist in development efforts through publicly coordinated service organizations such as the Peace Corps and AmeriCorps in the US.

This rising wave of people-centered, decentralized, often self-organized development, coordinated in large part by NGOs, represents a promising way to address sustainability problems and reduce spatial inequalities. However, it shouldn't substitute for official public sector efforts. Governments have far greater resources and capacity and can operate at the larger scales needed, for example, to address major climate justice problems and transition to a green economy.

Ramp Up Reparations

Reparations to particular groups of individuals, cities, regions, or countries may be appropriate as well. Ta-Nehisi Coates and others have argued that the US should pay reparations for both the slavery of African Americans and their systematic repression, disenfranchisement, and terrorization by whites in the time since slavery was ended.[36] Among the forms such reparations might take would be payments to cities with large African American populations to improve schools, services, health care, and housing. Individual members of oppressed groups might also receive subsidized home mortgages, money that could be used to support their education, or cash payments. Chicago adopted a reparations ordinance in 2015 providing $5.5 million in compensation to hundreds of African Americans tortured by its police between the 1970s and the 1990s.[37] Asheville, North Carolina, passed a reparations measure in 2020 to promote homeownership and business opportunities among African American residents.[38]

Other types of reparations might include payments from former colonial powers to the countries they colonized. Not all of the latter would qualify, of course; there is little case that Great Britain should provide reparations to the US and Canada. But many struggling, more recently separated nations in Africa, Latin America, the Middle East, and Asia could benefit from such direct assistance in recognition of the fact that historic subjugation has long-lasting effects.

Oppressed indigenous communities in most of the world's countries would qualify for reparations. For Sweden, Norway, and Finland this would mean the Sami people, formerly known as the Lapps, who lost land and resources during forced assimilation by the dominant Scandinavian cultures. For Thailand this would mean the multiple ethnic groups known as the "hill tribes," which inhabit the country's northern and western portions. For Mexicans, this would mean Indian communities in regions such as Chiapas. For the US, this would include Native

American tribes. Such recognition of past injustice could help indigenous communities materially as well as help prevent future forms of maltreatment.

Bring About Structural Change

Most important of all is rethinking the mechanisms that have created spatial disparities in the first place.

Although some degree of spatial inequality will always exist because of the geographic factors mentioned earlier, poorly regulated capitalism is a main driver of the disparities we see today. A basic feature of capitalism is its tendency to concentrate capital in the hands of some individuals rather than others, which then benefits certain communities and regions rather than others. The natural corrective would be to limit this process through progressive taxation, limits to the size of businesses, and the breaking up of monopolies. These strategies have not been sufficiently pursued because of the political power of wealthy interests.

The tendency of capitalism to export resources and profits from poor to rich places is another mechanism leading to spatial inequalities. Restrictions or steep taxes on financial transfers would be a way to address this. A company might be prohibited from transferring its profits from Brazil to the Netherlands, say, or might be able to move them only slowly, paying a sizable percentage back to Brazil to assist with that nation's development. Individuals might be restricted from transferring their wealth out of the country to bank accounts in Switzerland. We have already mentioned the possibility of a Tobin tax on global financial transfers. A 2002 study estimated that a tax of 0.1 percent would raise $76 billion annually, which could be used to fund the United Nations or be redistributed to disadvantaged societies worldwide.[39] Within the US, a 0.5 percent tax on stock transfers, as proposed by Senator Bernie Sanders in 2016, could raise between $34 and $340 billion (depending on how much this depresses stock trades), which could be used to assist disadvantaged communities and regions.[40]

Within a given country, legislation could require banks and other businesses to keep capital within communities rather than exporting it to other locations. The 1974 Community Reinvestment Act in the US does this to a modest extent, requiring commercial banks to lend to low- and moderate-income communities in which they do business. This act was intended as a first step to address historic redlining. However, it is a relatively weak program poorly enforced, and banks often lend to wealthier people in poor zip codes. A new version could require far more bank and business investment in disadvantaged communities, emphasizing activities that affect the least well-off residents.

Another type of structural reform could address the tendency of individuals and corporations under capitalism to accumulate profit while ignoring environmental and social externalities. Thus the coal industry avoids paying for many of its negative effects on Appalachia, the oil industry avoids assisting climate refugees, and tech companies ignore the way electronic waste from their products harms people in Asia and Africa, where much of the waste is eventually exported. Cities, states, countries, or regions could require firms doing business within their borders to greatly reduce these externalities, for example by taking life-cycle responsibility for the impacts of their products. The European Union began requiring businesses to do this in 2005. Firms could also be required to train local youth, give workers paid leave to engage in local service, provide affordable housing for their workforce, and hire a certain percentage of workers locally.

At the national scale, trade policy comes into play. Rather than pushing economic globalization, countries could limit long-distance trade on the grounds that it undermines local economies and exacerbates disparities. The European Union, for example, is famous for tariffs to protect its farmers and preserve its rural culture. At an extreme, India used import substitution policies (seeking to foster local industries for essential products rather than importing those products) after World War II to keep out entire categories of consumer goods such as cars and soft drinks in order to stimulate domestic manufacturing. To be effective, though, such strategies must be adopted by many countries. Otherwise overwhelming economic and political pressure is placed on the one or two societies trying to do something different, as happened with India.

The most difficult strategy to reduce spatial inequity, which would need to be undertaken in conjunction with any of the others above, is to change values and lifestyles. The high-tech business mantra of "Move fast and break things" may need to change to "Care for your neighbors and build a healthy world." Hyperindividualism, as emphasized by several generations of right-wing leaders worldwide who have drawn on social thinkers such as Ayn Rand, Friedrich Hayek, and Milton Friedman, has run its course. An approach of collective caring and responsibility is waiting to take its turn. Here, as with other topics, the Covid-19 pandemic helped expose the problems of current structural inequalities—some geographic communities suffered far more than others—and the need to address them.

CONCLUSION

Spatial disparities in economic wealth, community welfare, and environmental health have long existed. But the current inequities are truly enormous, both within countries and worldwide, and will only be exacerbated by the climate crisis. These disparities have downsides for both wealthy and disadvantaged cities and for the planet's ecosystems across multiple scales.

Sustainable cities will be ones in which people live locally and help care for other residents, communities, and ecosystems in their region. Cities will focus primarily on meeting local needs and collaborating with other cities on common goals, rather than competing with each other. Mutual assistance will be a way of life. Capital will remain local or regional rather than flowing to New York, London, Beijing, or Bentonville, Arkansas.

A more proactive public sector will ensure the same foundation of human rights and services within every community. Higher levels of government will help equalize financial disparities by redistributing funds from city to city and region to region and will coordinate reparations for past damage done to particular places and populations. Targeted assistance efforts by local, regional, and national governments, international organizations, NGOs, individuals, and the private sector will address the needs of disadvantaged communities. Structural reforms will change those aspects of capitalism that most generate spatial inequality and will develop alternative economic systems that do not concentrate wealth to such an extent and generate inequality. Businesses will be required to take responsibility for the environmental and social externalities of their operations.

Such a situation will be 180 degrees opposite to the individualistic, self-interested value system promoted by capitalism, populism, and nationalism. Ayn Rand, the mid-twentieth-century novelist beloved by neoliberal politicians, would not approve. But a new ethic of collaboration toward sustainable development goals will be essential to end the structural inequities plaguing cities and regions today and to prepare societies to address the climate crisis with its future additional burdens.

6

How Can We Get Where We Need to Go More Sustainably?

During the Covid-19 pandemic, millions of people worldwide changed their transportation habits. Individuals who had routinely flown across continents or commuted dozens of miles daily on congested roads suddenly found themselves working at home or not working at all. Many were surprised that they were able to accomplish so much remotely. Instead of spending time commuting, they had more time with their families. Some who had usually ridden buses or subways to work decided to bicycle instead and so got more exercise. Parents who had shuttled children from one activity to another no longer spent hours daily as family taxi drivers.

Meanwhile, after long days on Zoom, Google Meet, Teams, or WebEx, millions of people began walking regularly to get fresh air. With gyms and other athletic facilities closed, individuals who had never ridden before bought bikes. People suffering from social isolation adopted dogs for company and now took them for walks several times a day. But in neighborhoods that lacked green spaces and safe places to walk, residents felt trapped in their homes and frustrated. They suffered mental and physical health impacts.

Cities responded to these changing behaviors and people's need for socially distanced recreation. Oakland closed seventy-four miles of streets, 10 percent of the city's total, to through-traffic in order to allow pedestrians, cyclists, and playing children to use them.[1] Barcelona closed forty-four streets to motor vehicle traffic to allow pedestrian use, and

Bogotá created forty-seven miles of temporary bike lanes.[2] The Boston suburb of Brookline repurposed parking lanes as pedestrian walkways.[3] When restaurants were allowed to reopen, cities let those establishments put tables on roads, sidewalks, and parking lots outside their buildings to allow for more outdoor dining. In these and many other ways, the pandemic allowed cities to rethink transportation needs and ways to use urban space to meet human needs. Planners saw opportunities to reimagine.

The push toward sustainable cities in recent decades has begun to make cities more bike- and pedestrian-friendly, to reduce private motor vehicle use, and to lessen the externalities associated with transportation systems such as air pollution and GHG emissions. Cities worldwide have hired bicycle planners, developed "complete streets" (designed to accommodate bicycles, pedestrian, public transit, and green space elements as well as cars) and pedestrian trails, initiated new forms of public transit, and raised parking charges to discourage driving (figure 8). However, if we want to reimagine sustainable cities the shift toward more sustainable transportation systems needs to be taken to the next level and coordinated with better land use and urban design.

Existing transportation systems have enormous sustainability problems. Cars and trucks emit almost one-fifth of US GHG emissions, more than half of carbon monoxide and nitrogen oxide emissions, and a large share of hydrocarbon and particulate emissions.[4] About 165 million Americans live in areas where such air pollution has major negative health effects. In northern China air pollution takes five years off the average life span of some five hundred million people,[5] and worldwide it may take two years off the average life span.[6] More than forty thousand people die in vehicle accidents in the US each year, and hundreds of thousands worldwide.[7] Americans spend an average of 17,600 minutes apiece in their motor vehicles every year, the equivalent of thirty-six eight-hour days.[8] Imagine the benefit that would result if people spent this time with their families or in other more productive pursuits!

Reliance on private motor vehicles has created massive traffic congestion in many of the world's cities and has made the urban realm unpleasant, noisy, unhealthy, and dangerous. Many low-income individuals in particular are forced to commute long distances, spend a large proportion of their income on transportation, and suffer from pollution, noise, environmental injustice, congestion, and safety risks. Others who cannot afford private vehicles spend hours waiting for buses or trains on underfunded public transit systems.

FIGURE 8. Dunsmuir Street in Vancouver, B.C., has been redesigned as a "complete street" that gives ample space to bicycles, pedestrians, and traffic moving at a slow and steady pace. The street's design also includes trees, green spaces, and plazas opening to the side. (Paul Krueger/Creative Commons.)

Let us imagine instead cities where most people can bike, walk, or take public transit wherever they need to go; where households need at most one car and don't use it regularly; where children, women, the elderly, LGBTQ individuals and people of color can easily and safely travel anywhere without concerns for their safety; where sidewalks are in good condition and pedestrians are prioritized; where streets are greened and cooled by climate-appropriate trees and plants; and where outdoor spaces now dominated by motor vehicles are used by restaurants, cafes, performance artists, food trucks, markets, outdoor classrooms, and children rather than cars.

Such a vision is completely possible. We are all actors in the story of transportation, and we make daily choices about how to travel. Many aspects of transportation systems will need to be rethought and reimagined because of the climate crisis, and this is an opportunity to leverage many advantages for cities by changing our personal behavior as well as technologies, institutions, and infrastructure.

The solutions to sustainability challenges depend on how we frame the questions. One main reason transportation planning has been stuck in unsustainable modes for so long is that societies often do not ask the right questions. If, as has usually been the case in the past, planners wonder, "How do we meet expected transportation demand?," then transportation planning becomes a quest to add capacity in the form of new roads, bridges, and transit systems. These in turn often generate suburban sprawl and many negative externalities. Even if the question is "What might be more sustainable modes of transportation?," the discussion is likely to focus on the relative merits of different vehicle technologies, while avoiding more important topics such as how levels of driving can be reduced. If, on the other hand, we ask, "How can all people get to where they need to go sustainably?," a very different discussion is likely to emerge. The question of "needs" arises immediately and brings in lifestyle, urban design, environmental, and social equity considerations (Why are everyday destinations so far apart from each other? Why can't workers afford to live and raise families near their work? Do we really need to fly so many places so often?). Social justice questions arise, since clearly it is unfair that disadvantaged households should have to engage in unsustainable travel behaviors. Conversations around those topics are likely to lead to more sustainable and equitable solutions overall. This framing of the challenge forces us to think holistically and intersectionally about developing transportation for everyone.

With transportation, as with other sustainability challenges, solutions need to be tailored to context. For example, the State of California has begun building a high-speed rail system likely to cost at least $100 billion. On first glance that sounds great. Who could be against better public transit? High-speed trains operate successfully in Europe, Japan, China, Korea, and Morocco. However, the California context is very different from any of these, and even from the East Coast of the US. Apart from San Francisco, the state has few dense cities with good local public transit that would connect to a high-speed rail station. Ridership would probably be low. Train tickets would probably be expensive, meaning that many of the state's residents couldn't afford to ride the system. The more appropriate question given the context might be, "Is it a good idea to spend $100 billion to meet the needs of a relatively small number of affluent professionals traveling from San Francisco to Los Angeles when land use planning throughout the state doesn't support the system and reallocating the money to local buses and light rail would serve vastly more low-income people of color?" When reimagining sustainable cities, we especially need to ask about the equity implications of decisions. We do not have endless resources, so we need to prioritize investments using equity and sustainability as guides.

Transportation is a field in which "technological optimism"—the belief that technology will be the primary solution to problems—plays an outsized role. Focusing discussions on questions of technology leads us to assume that high levels of mobility are necessary and appropriate and that electric cars are the solution to everything. Although it may seem counterintuitive, a good framing for discussions might be "How might we *decrease* current levels of mobility but improve access and make it more equitable?" Cities like Los Angeles have lots of mobility—with its negative side effects (traffic, air pollution, traffic fatalities, etc.)—but poor access. To get anywhere, one must get on the freeway and sit in congestion, or else sit on a bus in traffic. And many of the bus routes require people to transfer and spend hours commuting. Angelenos' quality of life would be higher if each neighborhood had a good balance of local stores, restaurants, cafes, parks, schools, and workplaces that residents could walk to and that provided them with employment opportunities. But a century of poor development, oriented around private motor vehicle mobility, has meant that many parts of the region don't have those things. When we reimagine sustainable cities, we need to put those things back.

Recently experts have heralded multiple "revolutions" in the transportation field.[9] Electric vehicles are one, autonomous vehicles another, and shared ride services a third. Smartphone and Internet technologies potentially link such innovations together. Some authors propose that these technologies represent a new era of transportation that will provide "faster, smarter, greener mobility."[10]

Some of these innovations may indeed be useful for sustainability. Electric cars, for example, represent a main way to reduce GHG emissions. Car-share services may reduce the need for each household to own a vehicle. Yet these new technologies pose dangers as well. Autonomous vehicles, for example, may increase motor vehicle use if the occupant has a pleasanter experience watching television in the back seat instead of driving, or if vehicles circulate endlessly looking for parking after dropping off their riders. Major social equity questions exist around who will be able to afford these new technologies and if government investment in private car ownership is the most equitable use of government funds. It is also far from clear that such revolutions will be as quick or widespread as their proponents promise. Self-driving cars, for example, have been anticipated for years but have yet to be proven safe and useful in all conditions. When we put all our faith in technological fixes, we sometimes miss other opportunities such as those to strengthen public transit or promote more local and walkable economies.

Technical innovations are important but are unlikely to solve sustainability problems related to transportation. For that it is essential to ask good questions, understand contexts, and think in a holistic, intersectional way about human and ecological needs.

STRATEGIES FOR SUSTAINABLE TRANSPORTATION

At least five main sets of strategies are needed to answer the question "How might people get to where they need to go sustainably?" These strategies focus on rethinking how much we need to travel; changing land use and urban design; moving to electric and human-powered travel modes; changing incentives for travel; and ensuring transportation equity (table 6). All are interrelated.

Behavior Change

The starting point is to rethink how much we need to travel. The Covid-19 pandemic encouraged such reevaluation by confining many of us to our homes for extended periods. This was often challenging, to be sure, but many of us found that we really didn't need to travel quite as much as before.

The question for sustainability advocates is how to encourage more local lifestyles in which everyone satisfies daily needs and finds meaning in life closer to home. Behavior change relies in large part on changing the information people have available to them. Making sure a community's residents have and know about local parks, jobs, facilities, and cultural opportunities is a starting point. Celebrating those local opportunities and encouraging peer groups to use them is a further step. Schools can help children learn to appreciate their local community, local religious and civic organizations can take on local projects, and festivals and stewardship programs can engage residents in local activities.

Social marketing campaigns are at the heart of behavior change efforts, and pioneering municipalities have full-time staff developing programs of this sort. Vancouver, British Columbia, for example, developed an ambitious Transportation 2040 Plan in the mid-2010s aiming to have two-thirds of all trips in the city made by walking, cycling, and transit by 2040. As part of this plan, a working group of staff across city agencies developed an Active Transportation Promotion and Education

TABLE 6   STRATEGIES FOR MORE SUSTAINABLE TRANSPORTATION

Strategy	Description
Behavior Change	
Encourage reductions in personal travel.	Encourage people to work at home to the extent possible; live near their work; combine trips; take vacations close to home; and learn to appreciate local environments and community.
Promote local living.	Improve access to and information about local parks, jobs, facilities, and cultural opportunities; conduct social marketing campaigns and events aimed at promoting local lifestyles.
Promote car-free lifestyles.	Emphasize "active living" based on walking or biking; change codes to allow urban buildings in transit-oriented locations to be built without parking; offer economic incentives such as free transit passes and ride-share subsidies.
Encourage sharing of vehicles.	Require car-sharing pods within new development; make street parking space and public subsidies available for such vehicles; incentivize their use.
Land Use/Urban Design	
Promote more balanced land uses and neighborhood centers.	Change zoning codes and develop area plans that ensure a mix of land uses, green spaces, and housing types; encourage creation of neighborhood centers and corridors.
Require urban densities.	Revise codes and plans to require at least modest residential densities (~12 units/acre) so as to reduce community size and make bicycling, walking, and public transit use more feasible.
Promote "complete street" design.	Add bike lanes, sidewalks, and green spaces to streets; reduce unnecessary vehicle lanes and parking.
Electric and Human-Powered Travel Modes	
Require all-electric vehicles to end use of fossil fuels in transportation.	Set near deadlines for all-electric private vehicles and transit, since with renewably generated electricity this will greatly reduce GHGs and local air pollution.
Encourage biking and walking.	Use educational campaigns, urban design improvements, and economic incentives to encourage these modes, which are the best for short trips.
Improve public transit.	Increase public sector support so as to provide more frequent, better-networked public transit service for longer trips.
Encourage ride sharing.	Promote shared vehicles, especially carpooling with private vehicles, while ensuring that for-profit ride-sharing companies such as Uber and Lyft pay drivers fairly and do not destroy traditional taxi systems.

(continued)

TABLE 6 *(continued)*

Strategy	Description
Transportation Demand Management	
Implement true-cost pricing.	Use fuel taxes, tolls, parking charges, congestion charging, and vehicle registration fees to incorporate social and environmental costs of travel into the price the user pays. End free parking to encourage drivers to consider other travel modes.
Make public transit free.	Subsidize public transit heavily to reduce private vehicle use and improve equity.
Adopt place-oriented vehicle restrictions.	Implement toll rings, center-city pedestrian zones, and other steps to discourage private vehicle use in dense urban locations.
Improve information about alternatives.	Provide good information about public transit options, ride-sharing opportunities, bike route networks, and nearest available parking.
Transportation Equity	
Ensure inclusive communities.	Reduce travel distances for disadvantaged populations by requiring all jurisdictions to accept their fair share of a region's affordable housing, funding and upgrading affordable housing, prohibiting discrimination within housing markets, and incentivizing employers to locate jobs in disadvantaged neighborhoods.
Improve public transit.	Provide more frequent service, better-coordinated routes and transfers, better service on nights and weekends, and safer, cleaner vehicles and stops.
Make public transit cheap or free.	Provide financial incentives to reduce cost and increase ridership.
Ensure universal access to clean electric vehicles.	Make sure that lower-income households receive a proportionately greater subsidy, and ensure that charging stations are universally available.
Meet the needs of diverse populations.	Design transit, bike, and pedestrian systems to better meet the needs of differently abled individuals, seniors, children, parents with infants, and other groups.

Program. The group conducted market research on target audiences who walked or cycled occasionally but might do so more. They identified concerns about safety, confidence, and competency as behavior change barriers, and undertook pilot projects with community partners to address these. Many promotional events were held with specific groups such as parents of school-aged children. (The "walking school bus" is a specific strategy used in many cities to encourage children to walk to school again; parents walk around the neighborhood picking up kids along the way.) Vancouver's research also highlighted the importance of building an overall "brand" as a strong walking and bicycling city in order to encourage widespread behavior change. Copenhagen did the same with its "I Bike CPH" campaign.[11]

Social marketing campaigns can be launched around living locally, supporting local businesses, and living in more active and healthy ways—all of which help reduce travel needs. "Active living" campaigns, for example, have been pursued in many cities since the 2000s to help address the growing epidemic of obesity, with the support of the Robert Wood Johnson Foundation. These typically bring together coalitions of community organizations and health care providers to figure out strategies such as improving bike and trail networks, making streets more pedestrian friendly, and having physicians prescribe outdoor exercise to their patients.

Efforts to reduce individual travel can enlist economic arguments as well. Most people in the US would be surprised to know that they spend on average more than $8,000 a year owning a car.[12] Many workers would consider living closer to their work or walking, bicycling, or taking transit if they were offered a "parking cash-out" sum of, say, $60 a month for not using a parking space or if their public transit were very cheap or free.

Regulatory revisions can encourage behavior change. Many US jurisdictions currently require as much as one off-street parking space per bedroom within new development, a policy that promotes vehicle ownership and use as well as consuming large amounts of land and driving up housing costs. Changing zoning codes to allow car-free housing in locations near public transit and urban businesses can encourage reductions in motor vehicle ownership and use. Some individuals would be willing to give up their cars if their residences were within walking distance of most things they needed and/or they were ensured access to electrified car-sharing pods with vehicles that could be signed out whenever needed.

Lifestyle choices are generational in nature. Behavior shifts slowly, and change is often the result of younger generations deciding to live according to different values from their parents. There are mixed signals on whether transportation behaviors are already changing in this way. Millennials—born in the last two decades of the twentieth century—have often chosen to live in more urban locations than their parents, get their driver's licenses at a later age, and own fewer vehicles.[13] Some of this change may be because of reduced economic circumstances. Some of it may be because of a preference for spending more time at home,[14] perhaps the result of growing up with screens. Whatever the reason, this trend represents a shift toward more sustainable transportation behavior that can be supported by additional investments in affordable housing in cities and improvements and investments in urban amenities, public transportation, and public schools. It is likely that Generation Z (born in the early twenty-first century), which is already actively pushing for climate justice, will also be receptive to more sustainable lifestyles that minimize car dependence and GHG emissions. But we need to plan more equitable and livable urban communities where they can live full lives so they are not pushed out to suburban areas by the lure of good schools or green space or the lack of urban affordable housing.

Land Use and Urban Design

Good urban design can help bringing jobs, homes, schools, parks, stores, and other destinations of daily life closer together and make traveling between them without a private motor vehicle easier. One goal is to improve land use balance so that the destinations of daily life are close together and people can walk or bike to them. This means neighborhood centers or corridors throughout the city with an intensive mix of workplaces, stores, and institutional land uses such as health care, day care, schools, and government offices. These nodes and the surrounding residential areas must be dense enough to provide sufficient market for small businesses and to support safe and affordable public and nonmotorized transit. Related design goals are to create a well-connected street and path network so that walking and biking is possible and then to make streets and public spaces attractive and free from high-speed motor vehicle traffic. A green spaces network of parks, trails, and wildlife corridors, preferably along waterways, can create walking routes and an ecological armature for the community at the same time.

The development industry will not necessarily design neighborhoods in these ways, so the public sector must proactively plan out the urban fabric before the developers arrive, for example by establishing standards for street and path connectivity in addition to mapping out land uses and densities through equity- and sustainability-focused zoning. The movements known as the New Urbanism and Smart Growth have promoted such design strategies in recent decades.

Within existing cities, any large parcel of commercial or industrial land can be redeveloped as a walkable neighborhood with a grid of streets, small blocks, a mix of building types and uses, and green spaces. Any old strip mall can be reimagined in similar ways. Any suburban commercial street can be redesigned as a pedestrian-friendly boulevard with motor vehicle travel lanes at the center, planted median strips bounding those, bike-and-pedestrian spaces to the side, and three- to five-story urban buildings lining the edges. Any suburban town can proactively identify opportunities to add walkable neighborhood centers in these ways, allowing residents to live more locally.

The Central Park neighborhood within Denver, Colorado, provides an example of such community design. The former site of Denver's main Stapleton airport was transformed in the 2000s and 2010s into a mixed-use neighborhood with a commercial center and many forms of small-lot housing for seven thousand people. A well-developed green spaces system links fifty different public parks and green spaces, including restored waterways where airport runways once ran. The project is not perfect—local officials insisted on overly wide streets, for instance— but represents the type of infill development that in the long run can help reduce motor vehicle needs in cities.[15]

Electric and Human-Powered Travel Modes

As pedestrian advocates have argued for generations, the most important transportation modes for sustainability are those that are human-powered. Walking is most crucial, followed by use of bicycles, scooters, skateboards, inline skates, and other human-powered conveyances. However, large corporations are unlikely to promote these modes, since there is often little money to be made from them, so the public sector needs to facilitate their use in cities.

Cities can prepare bicycle and pedestrian master plans identifying priority route systems and context-appropriate design strategies. They can reconstruct roads as "complete streets": this might mean, for example, taking a four-lane arterial and narrowing it to one motor vehicle lane each way with turn pockets at intersections, adding bike lanes, planting strips, and wider sidewalks, depending on the right-of-way width. (The names "open streets" and "slow streets" are also used for roads primarily intended for nonmotorized use.)

Cities can also create off-road paths for cyclists and walkers, including large-scale bike/ped bridges over freeways, railroad tracks, and waterways such as European countries are now building. Copenhagen, for example, built seventeen new bike bridges over canals and waterways as part of a comprehensive bicycle infrastructure plan that helped raise bicycle commuting to work or school from 36 percent in 2012 to 62 percent in 2019.[16] Electric bikes, scooters, and skateboards are also gaining in popularity as low-tech alternatives to motor vehicles, although they may also conflict at times with bikes and pedestrians. In some cities, bike-sharing and scooter-sharing programs are increasing access and mobility. These programs can be designed in conjunction with public transit investments to promote equity.

Many forms of public transit will be needed within the sustainable cities of the future, which means that we need to invest in it. Light rail (streetcars or relatively short trains running primarily on city streets) and heavy rail (metro systems running on their own tracks underground or away from streets) technologies have been expanded in many urban regions worldwide since the 1970s. These can potentially accommodate very large numbers of passengers in a nearly GHG-free manner if using renewably generated electricity. In the post-Covid-19 world, such transit modes face new hurdles of public acceptance. Upgraded sanitation procedures, ventilation, and opportunities for social distancing are needed. Drivers especially need to be protected, since they are exposed to members of the public all day. The good news is that buses, subways, and trolleys appear not to have been responsible for many coronavirus cases. In Paris, researchers found that none of 386 Covid-19 infection clusters were linked to public transit.[17] Studies in Austria and Japan found similar results. However, it remains to be seen whether all riders can be attracted back.

Bus networks are often neglected in North America in comparison with rail systems, which tend to attract wealthier riders. This is unfortunate, as bus systems are typically far more cost effective and extensive than rail systems, and low-income riders rely on them to get to work. The Los Angeles transit system, for example, consists of ninety-three rail stations but almost fourteen thousand bus stops.[18] Electrifying buses, improving headways (time intervals between buses), adding more buses to reduce crowding, better coordinating connections, increasing sanitation, and expanding late-night service are all ways to upgrade this mode.

Bus rapid transit (BRT) is a form that has spread rapidly worldwide in the past generation. In these systems, articulated or biarticulated buses usually operate on their own lanes with widely spaced stops, improved stop design to speed loading and enhance rider comfort, and ability to preempt traffic signals to improve speed. The idea is to emulate the speed, quality, and experience of light rail service at a fraction of the cost. Curitiba, Brazil, famously had one of the first such systems, designed with its own rapid-boarding tubes and dedicated rights-of-way down the middle of arterial streets. BRT systems have now spread to hundreds of other cities worldwide, including Bogotá, Kampala, Lagos, Beijing, Guangzhou, Los Angeles, Mexico City, Quito, and Helsinki. Both BRT and regular bus systems can be converted to run on renewably generated electricity so as to be zero-emissions for their operations. Hydrogen fuel cells can also be useful for buses, which can easily accommodate large fuel tanks along the top of the vehicle, assuming the hydrogen is produced using renewably generated electricity.

Dial-a-ride services such as Uber and Lyft (and Didi Chuxing in China, Grab and GoJek in Southeast Asia, Careem in the Middle East, and EasyTaxi in Latin America) could potentially usher in a new era of shared mobility in which people simply call rides on demand rather than owning vehicles themselves. But major questions exist around social equity and environmental impacts. Whether they will increase or reduce vehicle miles traveled (VMT)—the basic metric of motor vehicle use—and road congestion is still uncertain. These services also decrease ridership on public transit systems, put traditional taxis out of business, and reduce political support for transit investment. Drivers are often poorly paid and in effect become part of a new underclass in societies serving the wealthy.

Electric vehicles are the future, for both shared and private travel, as this is the way to reduce transportation GHG emissions the fastest if the electricity is renewably generated. Already electric cars and trucks are spreading worldwide, helped by batteries that provide significant range (~250 miles) before recharging. Larger lithium batteries can allow a 500- to 600-mile range, and a new generation of solid-state battery technologies in the years ahead will provide even greater range with recharge times of under ten minutes.[19] The single most important step jurisdictions can take regarding transportation sustainability is to require all vehicles sold to be electric. California and many other jurisdictions have committed to doing this starting in 2035 or sooner. National and international leadership on electric vehicles will also help speed this much-needed transition.

Hydrogen fuel-cell vehicles also can be zero-emissions if using hydrogen produced with renewably generated electricity. But hydrogen is a more cumbersome fuel less likely to succeed in the small-vehicle market. Plug-in hybrid vehicles are a relatively sustainable transportation solution as well, since they provide a small range (typically ~20–30 miles) in all-electric mode before gas engines kick in, and yield overall equivalent mileage of one hundred miles per gallon of gas. However, since the need is to end fossil fuel use altogether, hybrids are best seen as a transition technology.

Other public policies such as rebates, tax credits, and publicly provided charging stations can help electric vehicles replace gasoline-powered ones. However, simply converting to electric vehicles will not make current patterns of transportation sustainable. Electric grids will not be 100 percent renewable anytime soon. There are also significant embodied energy and emissions in electric vehicles themselves, and they will still produce traffic congestion, accidents, and noise (from tires if not engines). People will still lose precious hours commuting, and continued heavy use of private motor vehicles will still fuel suburban sprawl unless we reimagine our relationship with mobility.

High-speed rail, referred to toward the beginning of the chapter, is often promoted as a more sustainable alternative to air travel. In urban regions with compact, high-density cities around stations it may be this. However, its impacts in terms of GHGs are still substantial—lots of electricity is required, and only in places like Sweden with its hydroelectric resources and France with its aging nuclear power plants does electricity currently come primarily from non-climate-affecting sources. Track, stations, and rolling stock contain large embodied GHG emissions, and noise pollution near tracks can be high. Equity questions exist concerning who can afford it. So it must be explored with caution, and reducing the frequency and necessity of long-distance travel altogether is likely to be the more sustainable approach.

Air travel is the most problematic transportation mode of all. Jet aircraft emit enormous volumes of CO_2 and water vapor (itself a powerful GHG). No battery-powered technology exists or is likely to anytime soon for long-distance, high-capacity aircraft, given the weight of batteries. Worldwide, air travel is one of the most rapidly growing sources of GHG emissions. One radical proposal that may become necessary is to ration everyone's flights. As proposed by British writer Sonia Sodha, each person could be given a certain air travel allowance annually, probably no more than one or two trips. If individuals choose not to use their allowance, they could sell it through a publicly regulated market. Others who wanted or needed to travel more could buy them. This framework would incentivize local lifestyles and business use of teleconferencing, while providing a handy source of income for poor individuals in need of it.[20] During the Covid-19 pandemic the need for much business travel was brought into question. Quite likely video links can serve in many cases instead, and businesses can reinvest their travel budgets in other things like paying their employees living wages.

Transportation Demand Management

Economic, informational, and regulatory strategies represent a broad set of tools to reduce demand for transportation, especially private motor vehicle use, thus improving urban sustainability. They are often referred to as Transportation Demand Management (TDM) policies. As with energy consumption, a variety of "carrots and sticks" can be employed in this way.

Fuel taxes or carbon taxes—which amount to much the same thing for fossil fuel–powered vehicles—provide both a carrot (by incentivizing less driving and more efficient vehicles) and a stick (by raising costs with every fill-up). Many countries, especially in Europe, currently use these mechanisms. Taxes range from 0.78€ per liter ($3.28/gallon) in the Netherlands and 0.73€ per liter ($3.07 per gallon) in Italy at the high end of the European spectrum, to 0.36€ per liter ($1.51 per gallon) in Bulgaria at the low end.[21] US federal gas taxes are $0.18 per gallon; states add an average of $0.34. These rates will need to be raised, particularly as vehicle fuel efficiency goes up and the wear and tear on the roads remains the same.

However, raising gas taxes alone doesn't greatly reduce driving. Demand for gasoline is relatively inelastic with respect to gas prices. Another problem with fuel taxes is that they are regressive (poor individuals pay a higher percentage of their income than wealthy ones). Plus they are highly visible and lead to political resistance. Reaction against gas taxes was part of the motivation in France in 2018 for the Yellow Vest movement.[22] Carbon taxes are likewise regressive unless redistributed to people in the form of a dividend so as to be value neutral, or refunded disproportionately to lower-income households.

Parking charges are often more effective than gas taxes at reducing vehicle use and attract less opposition. One important strategy is to get commuters to pay the true cost of "free" parking in lots they use every day, rather than have companies provide this parking as an unacknowledged benefit for workers who drive.[23] Making the cost of driving visible in this way can change behavior. When the Bill & Melinda Gates Foundation opened a new $500 million headquarters in Seattle in 2008, the city made it charge employees $12 daily for their parking. Employees also got free transit passes, bike lockers, and showers for bike commuters, plus a $3/day incentive for not driving. Whereas 90 percent of employees originally drove alone to work, a year later only 42 percent did, and eight years later the figure had fallen to 34 percent. Fewer than half of the parking spaces in the foundation's garage were used.[24]

Yet another economic strategy for reducing motor vehicle use is represented by congestion charging systems in cities such as Oslo, Stockholm, and London. First implemented in 2003, London's Congestion Charge is aimed at reducing the number of people who drive cars to central London. To do this, the city created a Congestion Tax of £8, later raised to £11.50, which drivers are automatically assessed when their vehicle enters the central city. The revenue is used to improve the public transit system, demonstrating that some sustainability strategies can subsidize others. Private vehicle traffic into central London fell 39 percent in the system's first decade. Unfortunately, for-hire vehicles including Uber and Lyft were exempt from the charge, leading to resurgent congestion.[25]

London also disincentivizes polluting vehicles in the city center with an Ultra Low Emission Zone (ULEZ). It developed this policy because it was found that about two million people, four hundred thousand children, and more than four hundred schools were located in parts of the city with "illegal pollution levels."[26] Cars that do not meet low emissions standards are charged an additional £12.50 on top of the Congestion Charge. The city also provides financial incentives for drivers with polluting cars and trucks to scrap their old vehicles.[27] To address the burden that might be placed on small businesses, London has set aside £48 million to help businesses upgrade to cleaner vehicles.

Paris has used an even broader set of policies to reduce vehicle use. Successive mayors have promoted safer pedestrian spaces by widening sidewalks, eliminating parking, adding bike lanes, and improving green space. In the 2010s, the city created six hundred miles of bike lanes and closed parts of the Seine River to car traffic on the weekends. It also started a bike share program called Velib, banned older cars on weekdays, made some major squares car-free, and created dedicated BRT lanes. As a result of these improvements, the automobile mode share has dropped 45 percent between 1990 and 2018 within city limits while transit use increased by 30 percent and the bicycling share rose tenfold.[28] Traffic fatalities were reduced by about 40 percent. In 2020, diesel cars built before 2010 were banned from the city.[29] Suburban jurisdictions adopted a similar ban beginning in 2025.

Other regulatory strategies can potentially discourage excessive motor vehicle use. In 2011 Beijing started a lottery for new license plates because of the city's extreme air pollution levels. The goal was to cap the number of gas-powered vehicles in Beijing at 6.2 million, for a city with 22 million people. Electric vehicles received preference under the system. In addition, even winners of the license plate lottery could not use their cars one day of the week.[30] Other cities have attempted similar policies. Mexico City adopted a *Hoy No Circula* policy under which cars were not allowed to drive on the basis of whether their license plate ended in an odd or even number. However, such strategies can backfire. The impact of Mexico City's policy was that many wealthy people bought a second (often older and more polluting) car to use on the alternate days. The lesson is that policies need to be carefully evaluated and adapted to ensure that they are achieving their sustainability goals and are not leading to unintended negative consequences.

Informational strategies, in addition to the local living campaigns mentioned above, are also part of TDM frameworks. Good information about public transit options and ride-sharing opportunities can help reduce the percentage of people driving their own vehicles. Well-publicized bike route networks can encourage cycling. Information about the health benefits of bicycling and walking can entice people to try those modes of travel. LED displays with real-time information about which municipal garages still contain available parking spaces can help reduce the number of motorists circling blocks in busy downtowns, reducing in small but important ways street congestion and air pollution.

TRANSPORTATION EQUITY

In metropolitan regions where people remain segregated by class, race, religion, and/or ethnicity, disadvantaged groups need to travel at unsustainable levels to reach workplaces and other essential destinations. This is profoundly unfair for them—requiring increased expenditures of both time and money—and increases traffic congestion, GHG emissions, and local air pollution.

Careful planning and design can promote transportation equity. Most fundamentally, people from disadvantaged groups need to be able to live in the same neighborhoods and jurisdictions as more privileged individuals. National and state policy can combat segregation by requiring and funding affordable housing in every community across metropolitan regions. Laws prohibiting discrimination in home mortgage lending, home insurance, landlord rental practices, and the real estate industry can be strengthened and rigorously enforced. Decades of societal disinvestment in communities of color can be reversed through incentives for employers to locate jobs in those neighborhoods, thus reducing resident transportation needs. Only when such steps have helped create more diverse and inclusive communities will the unequal transportation demands faced by different demographic groups be ended.

Cost is a barrier often preventing lower-income people from using public transit. However, creative policy initiatives can reduce this obstacle. In Philadelphia, the public transit agency, SEPTA, traditionally required children over five to pay the adult fare. With multiple children it became more cost effective for families to drive to destinations and park (even paying for parking) or not travel at all. Recognizing the equity impacts of this policy, SEPTA changed it in 2020 so that all children under twelve accompanied by a paying adult ride free.[31] This step recognizes that families are a planning and traveling unit.

Kansas City, Missouri, went further that same year, initiating free bus service for residents. The move costs the city $8 to $9 million annually but makes it much easier for low-income residents to get around. Other US cities are considering a similar move, making public transit free. Between 1975 and 2012, Portland, Oregon, maintained a "Fareless Square" policy in which bus and rail rides were free in the downtown area. Unfortunately, the transit agency had to end Fareless Square because of budget shortfalls following the Great Recession. International cities such as Tallinn, Estonia, and Dunkirk, France, also have free transit policies.[32]

Bus rider movements have arisen at times to push for more equitable transit investment. The Los Angeles Bus Riders Union, representing four hundred thousand Los Angeles region bus riders, argued in the 1990s that the Los Angeles County Metropolitan Transportation Authority violated civil rights laws because the bus system, which primarily served low-income people and people of color, received significantly less funding than the regional rail, which catered to a higher-income, more white population.[33] The Bus Riders Union won an important victory in the form of a legal settlement reallocating $2.7 billion to improve the bus system to address equity concerns.[34]

Improving transportation equity means attending to ways in which current systems ignore the needs of diverse populations. In New York City, for instance, accessibility groups have been suing the city for years to get elevators installed at subway stations. Their installation will make riding public transportation an option for many passengers with disabilities. Accessibility provisions are also critical for baby strollers, families with young children, people with medical conditions, and the elderly.

Another problem is that many low-income employees work night shifts or swing shifts and so require transportation to their workplace at odd hours. Yet many public transit systems stop running after midnight or have very limited service. Some level of round-the-clock service is essential to serve these workers. Many other transit riders face complex, slow, expensive, and awkward transfers between different transit routes. Steps to streamline transfers and improve headways (time between buses or trains) are often called for.

Bicycle planning is often focused mainly on able-bodied, healthy young adults. This is an important demographic. However, it is also important to plan facilities friendly to children, the elderly, those with lower levels of physical ability, and those who may never have ridden a bike before. This is where e-bikes may come in, but having safe, grade-separated bike lanes, separate bike crossing phases at traffic signals, bike underpasses or overpasses to take lanes over busy roads, and safe, secure bike parking near destinations can also help.

As fossil fuel–powered cars and trucks are replaced by electric ones, access to the new technologies has become an equity issue. Up-front costs of electric vehicles are an obstacle for low-income households. Existing incentives for purchase of such vehicles have often gone to affluent buyers who can afford the initial investment. Making those subsidies inversely proportional to income—very high for low-income households and not available to the wealthy—is one potential strategy for the future. Making electric vehicle chargers widely available to low-wealth as well as wealthy households, and renters as well as homeowners, is another step likely to be necessary.

In these and other ways, current transportation inequities can be greatly reduced. Doing so means foregrounding equity in all dimensions of urban planning and design. Simply improving public transportation or providing subsidized transit passes is not enough if disadvantaged groups continue to be geographically segregated. A more truly inclusive society is required instead that centers equity and then designs cities to meet people's diverse needs.

CONCLUSION

The twentieth century most likely represented the peak of emphasis on unfettered mobility. However, by designing low-density, spread-out communities based on this principle, developers ironically reduced residents' access to many places they need to go in daily life. They also created terrible environmental and social externalities. In the twenty-first century, if we plan with sustainability, equity, and racial justice in mind, the public may see less mobility but more access. Each of us will need to travel less when we plan our cities and regions better. That is hours of our lives returned to us, where we are not trapped in our cars and waiting in traffic.

In the sustainable city of the future, people will be able to get to places they need to go by a much greater range of means, none based on fossil fuels or directly producing GHGs, and many of them healthier and more social. Individuals will run into friends and neighbors walking, biking, or taking shared forms of transit. They will get to know their own neighborhoods far better and will help make them better places. Mobility will be shared far more equally within societies, with everyone having access to a wide range of options and no one sequestered away in poverty or else inside private limousines or jets. The advantages to the environment of such a future will be enormous, and the benefits in terms of healthy individuals and healthy societies will be just as large.

7

How Do We Manage Land
More Sustainably?

As more than one-quarter of a billion people have migrated from its countryside to cities in the past generation, China has developed its urban
areas more rapidly and on a larger scale than any other society in history.
But its patterns of land development unfortunately mirror common problems worldwide. Piecemeal development of high-rise apartment buildings,
factories, and expressways covers vast stretches of farmland without the
land use balance or connected street systems of cities developed in previous eras. New commercial businesses line outlying roads. Fences and
gates separate the rich in their green, affluent enclaves from everyone else.
At the center of cities, old neighborhoods (known as *hutong*s in the country's north) are pulled down, creating huge piles of rubble and depriving
the country of much of its architectural heritage. Wetlands are filled, villages razed, farmlands lost, and millions of people displaced.

Similar stories play out worldwide. Modern land development over
the last century has been a disaster for sustainability and desperately
needs to be reimagined. After all, land is one of our most precious
resources. How societies use it will determine their future access to
food, water, and air, as well as their ability to reduce GHG emissions.

Let us imagine future cities where land is used efficiently rather than
covered with poorly connected subdivisions, enormous highways, office
parks, auto malls, and fast-food outlets; where firm lines separate the
city from country, and suburban sprawl no longer consumes farms, forests, wetlands, and deserts; where workplaces and stores are mixed with

housing so that no one needs to travel far to get to them; where the main determinant of land use is not short-term financial gain but rather long-term ecological and social well-being (figure 9).

How might societies manage land more sustainably in such ways? As with other sustainability questions, there is no single answer. But there is a rich history of land management strategies worldwide to draw from, and an urgent need to brainstorm new approaches for the future.

EVOLVING ATTITUDES TOWARD LAND

Before the modern era, people often managed land collectively and valued it for spiritual, cultural, ecological, or social reasons apart from its financial value. Some indigenous and non-Western cultures still do this. Behavioral norms and traditions governed shared use of land, and management strategies often sought to protect common-pool resources, share labor, and preserve ecological systems. At least one-third of the land in medieval Britain was held in common and managed by local residents, who used it for pasture, agriculture, or hunting.[1] Communities in parts of China traditionally protected forests and sought to preserve species diversity by not overusing any one type of plant.[2] Taboos, plant worship, and moral restrictions enforced this behavior. Small-scale farmers in places as diverse as Nepal, the Philippines, and New Mexico managed irrigation systems collectively to distribute water equitably, share labor, and ensure continued agricultural production.[3]

Having land and resources managed through economic markets is a relatively recent invention, accompanying the rise of capitalism in the seventeenth and eighteenth centuries.[4] Throughout history individuals or groups at times controlled large amounts of land through military, religious, imperial, colonial, or feudal systems. But frequent buying and selling of land is more recent, and large-scale, publicly traded corporations specializing in land development date only to the twentieth century. The new types of land development businesses include "production home-building" companies constructing hundreds or thousands of units at once in large subdivisions, urban development corporations building new districts within cities worldwide, and real estate investment trusts (REITs), which own vast portfolios of apartment complexes, shopping centers, and office parks.[5] New types of small-scale land developers have also arisen. These range from mom-and-pop "flippers" who fix up and resell dilapidated properties to nonprofit developers building affordable housing in places where the market has failed to do so.

Land is now valued most fundamentally for its exchange value rather than its use value. Land markets are dynamic and global. "Growth machines"—political coalitions of landowners, developers, construction companies, real estate professionals, and construction workers who together push for rapid land development—dominate local politics in many places.[6] Libertarianism promotes the belief that "I should be able to do what I want with my land." Ethics of preserving land in a natural state, managing it for collective benefit, or revering and worshipping it seem far away.

However, even in capitalist countries various public rights over private property still exist. For example, in much of northern Europe the ancient "freedom to roam," which the Swedes call *allemansrätten* ("every man's right"), still exists in modern law. This gives members of the public the right to travel across private property. Rights to this effect still exist in the Scandinavian countries, the Baltic nations, Iceland, Scotland, Belarus, Austria, the Czech Republic, and Switzerland.[7] The right of access is usually on foot rather than by motor vehicle, and those using private property are expected to respect the land and its inhabitants.

More broadly, the rise of urban planning, zoning, subdivision, and building regulations beginning in the twentieth century added a large set of public constraints on private land ownership. Owners in most places cannot subdivide and sell as buildable lots pieces of land that lack water, road access, or other crucial services. Rather, they must go through a formal subdivision process so that the public sector can ensure compliance with basic social norms and environmental protections. Likewise, owners cannot build anything they want but must comply with local zoning and building codes. Granted, these may be weak or poorly enforced in places.

These three types of ordinances have brought some order to the land development process and ensured safer, more decent housing, but they are relatively weak tools for regulating large-scale patterns of urbanization, since they apply mainly to individual parcels of land, often ignore ecological considerations, and can be easily overridden by local officials. Societies began experimenting with larger-scale growth management tools during the late nineteenth and early twentieth centuries. German states empowered municipalities to regulate development, expropriate land easily if needed for public purposes, provide housing, and limit private profits from land speculation.[8] Dutch governments gained similar powers, with more emphasis placed at the national level (the Netherlands being a small country), and a strong role for

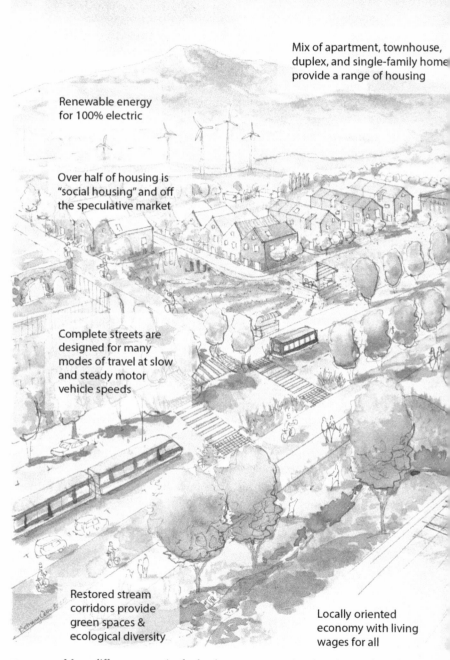

Mix of apartment, townhouse, duplex, and single-family home provide a range of housing

Renewable energy for 100% electric

Over half of housing is "social housing" and off the speculative market

Complete streets are designed for many modes of travel at slow and steady motor vehicle speeds

Restored stream corridors provide green spaces & ecological diversity

Locally oriented economy with living wages for all

FIGURE 9. Many different strategies for land use, transportation, housing, ecological restoration and other needs must fit together to create more sustainable types of urban form. (Bethany Celi

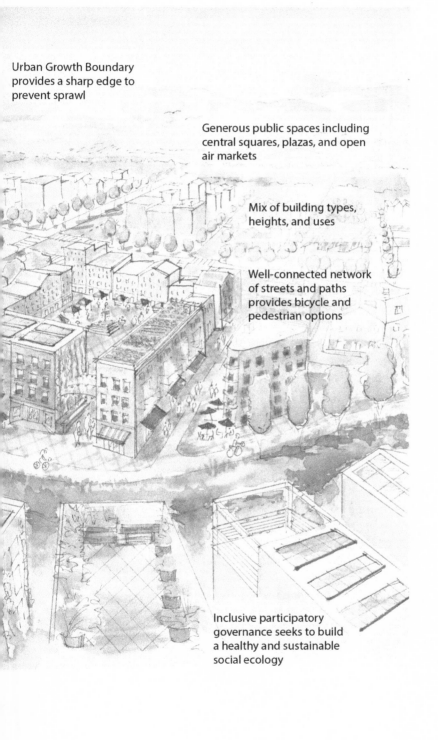

Urban Growth Boundary provides a sharp edge to prevent sprawl

Generous public spaces including central squares, plazas, and open air markets

Mix of building types, heights, and uses

Well-connected network of streets and paths provides bicycle and pedestrian options

Inclusive participatory governance seeks to build a healthy and sustainable social ecology cooperatives in developing housing. Many northern European nations used municipal land banking aggressively to manage urban expansion in the post–World War II period. Under this powerful mechanism for asserting public control over land development, a municipal agency owned most land on the urban fringe, leasing it to developers at times and places that made sense, and thus recapturing much of the increased land value arising from public investment in infrastructure such as transit systems. Cities such as Stockholm and Helsinki have controlled more than 60 percent of metropolitan land in this way since the mid-twentieth century.[9] In the post–World War II period, many northern European nations also developed strong regional planning to manage development around their major cities. Although public control over land has faced a severe backlash from private landowners in many countries during the neoliberal era, it remains necessary if farmland and important ecosystems are to be preserved, and if urban development that minimizes GHGs and best promotes sustainable living is to be undertaken.

New, more ecocentric approaches to land pose an additional challenge to the market's dominance. Environmental philosopher Aldo Leopold's "land ethic," proposed in 1949, states that "a thing is right when it tends to preserve the integrity, stability, and beauty of the biotic community. It is wrong when it tends otherwise."[10] Leopold's ecocentric ethic puts the well-being of the land and nature ahead of individual human interests, signaling a new balance between human use and ecosystem well-being. This philosophy would allow legal rights to be established, for example, for rivers to run unimpeded and for species to continue to exist. Already certain nations are taking steps in those directions. New Zealand, India, and Ecuador, for instance, have granted legal rights to major rivers within their jurisdictions.[11]

Leopold's land ethic could potentially be expanded in nonecological dimensions as well. A given parcel of urban land may be flat, devoid of wildlife habitat, and generally uninteresting. But it still occupies a unique spatial location within a city or town. It sits within a nexus of real or potential human relationships. These have economic, social, cultural, class, racial, and spiritual dimensions. What happens on this piece of land matters in many different ways to many different people. Development on it can potentially build community, add beauty, heal social rifts, and improve ecosystem function. Or these opportunities can be missed, and damage to the social ecology done instead. The person who controls the land can thus be seen to have a responsibility to use the land well—a "social ethic of land use." This extension of Leopold's principle could and should guide urban planning, design, and development decisions.

Although necessary at times, public regulation of land has run into problems in many countries. Especially in North America, zoning codes and development standards came to enforce a motor vehicle–oriented style of suburban development with spread-out housing, segregated land uses, wide streets, and massive amounts of parking. Codes grew to be hundreds of pages long and often regulated the wrong things, for example allowable activities inside buildings, when what matters most to local residents is the external look and feel of particular sorts of development, as well as ecological effects and impacts such as traffic and GHG emissions. Zoning and subdivision codes have segregated land uses from one another, placing vast amounts of single-family housing in some neighborhoods, while apartments, jobs, and commercial facilities are elsewhere. The resulting metropolitan landscape is highly fragmented, unequal, and dependent on fossil fuel–guzzling motor vehicles.

In the US, with its emphasis on decentralized authority to regulate land, thousands of local governments make land use decisions that add up to suburban sprawl, with its negative environmental and social equity externalities.[12] Redlining (financial institutions' refusal to make loans for housing in African American neighborhoods), racially restrictive covenants (deed restrictions preventing nonwhite or Jewish people from buying houses in many neighborhoods), exclusionary zoning (allowing only large, expensive lots that poor individuals could not afford), and federal investments in suburban-serving highways have served as additional mechanisms propelling suburban sprawl, contributing to the Black-white wealth gap, and exacerbating regional inequality.

MORE SUSTAINABLE LAND MANAGEMENT STRATEGIES

It is time for a wholesale rethinking of strategies to manage land. Specific new tools are needed as well as new values governing how societies and individuals use land and prepare for climate challenges. Some places have done relatively well at meeting goals such as preserving farmland or limiting concentration of land ownership in a few hands, and much can be learned from them. However, since no place is perfect and sustainability strategies must be tailored to particular contexts, a large toolbox of potential policies and programs is needed, along with a reimagining of the role of land use planning (table 7).

TABLE 7   STRATEGIES FOR MORE SUSTAINABLE LAND MANAGEMENT

Strategy	Description
Strengthen growth management planning and regional governance.	Embrace high-level government role in setting land management priorities; give regional agencies at metropolitan, watershed, or bioregional scales power to coordinate local action.
Adopt better forms of development regulation.	Develop simpler, clearer forms of regulation than traditional zoning, for example form-based codes, performance codes, generative codes, and ecologically based codes.
Promote compact, connected, well-balanced land development.	Limit outward sprawl through regional planning, urban growth boundaries, land banking, conservation easements, and provisions to facilitate infill development.
Encourage finer-grain land development.	Establish maximum building, block, and subdivision sizes; promote smaller-scale, more incremental land development within a publicly established urban design and planning framework.
Preserve ecological function during land development.	Protect waterways, wetlands, floodplains, wildlife corridors, and significant patches of habitat; actively plan and offer incentives to restore and expand these features.
Set maximum limits to land ownership.	Establish maximum amounts of land one person can own; redistribute land; limit absentee ownership of land.
Recapture speculative gains in land value.	Tax land rather than buildings (Henry George); adopt taxes on the "unearned increment" in land value, especially when resulting from public investment; adopt antiflipping taxes while encouraging long-term investment in urban property.
Expand collective ownership and management of land.	Revive the "commons" through institutions and incentives for common-pool management; use land banks to expand public ownership of land; establish co-ops and community land trusts.

Strengthen Growth Management Planning and Regional Governance

Establishing policies about land—in particular regarding urban growth management—is best done at high levels of government, since at the local level parochial interests of landowners and developers tend to have great political power. So national, state, provincial, and regional land management policies are particularly important. National land policy has often taken the form of establishing national park systems, wildlife refuges, forests, and seashores. But high-level policy for urban development and growth management (in Europe often called spatial policy) is just as important and often missing.

Country-level initiatives have been responsible for some of the most noteworthy urban planning initiatives historically. These include the Dutch "Green Heart" strategy, which since the middle of the twentieth century has preserved agricultural land in the Netherlands within a ring of cities including Amsterdam, Utrecht, the Hague, and Rotterdam.[13] They also include the creation of greenbelts and new towns around large British cities, especially London. The 1944 Abercrombie Plan helped established both the London Greenbelt and the first wave of British new towns, intended to reduce overcrowding in the central city. The postwar British government eventually constructed thirty new towns throughout the UK housing two million people, turning a profit in doing so. Other countries including Russia, France, India, Singapore, and China have pursued new town programs as well, with mixed results. However, if done well such high-level spatial planning can potentially preserve farmland and open space, provide high-quality housing for people from a mixture of social strata, ensure transit-oriented development, and help resettle populations displaced by the climate crisis.[14]

During the second half of the twentieth century, Japan, France, Germany, and the Scandinavian countries also pursued strong policies to preserve farmland. South Korea established greenbelts around its cities in 1971 following the London model, although it softened development controls in the 2000s under pressure from landowners.[15] Such urban containment strategies have at times been blamed, particularly in the US, for increasing housing costs within cities. However, aggressive public sector efforts to promote infill development, provide affordable housing, or—following the example of Britain—build new towns elsewhere could anticipate, plan for, and avoid this problem.

Growth management planning at the regional scale—meaning the scale of a metropolitan region, a bioregion, or a watershed—is essential for a sustainable urban future. Unfortunately this is a neglected scale of action in many parts of the world. Typically metropolitan regions are broken up into a central city and numerous suburban jurisdictions, which often do not work together or share resources. In fact, they often compete and replicate similar services while losing economies of scale. Agencies that plan for watersheds or bioregions are also rare and often underresourced or advisory.

Relatively strong regional planning has existed historically in France for the Paris region, which is central to the French national identity. Among other things, this has resulted in an excellent, multilevel regional transit system, and a number of less successful high-density modernist housing developments, known colloquially as *banlieus,* on the outskirts of the region. In Denmark, regional planning is responsible for the "finger plan" of development along rail lines extending from Copenhagen. Established in 1947, this vision has helped cluster development in transit-accessible locations and preserve green wedges of farmland and open space between these five "fingers" emanating from the central city. Instead of letting private speculators capture the increase in land value around rail stations, public agencies such as the City and Harbour Development Corporation recycled much of this revenue to help pay for transit improvements.[16]

In 1975, the Vancouver, British Columbia, regional planning agency (now known as Metro Vancouver) first prepared a Livable Region plan, which sought to create four regional town centers and to protect green spaces throughout the region. This plan helped catalyze dense but green redevelopment of the former industrial area around the False Creek bay to the south of downtown, creating a network of waterfront parks and trails in addition to slender, high-rise residential housing. In 1996 the agency prepared a Livable Region Strategic Plan that added two additional regional centers and placed a strong emphasis on sustainability and particular features such as regionwide bicycle and pedestrian facilities. A Metro Vancouver 2040 vision in 2011 then sought to plan for an additional one million regional residents by containing urban growth, channeling it into existing urban centers, and promoting more balanced "complete communities" that could reduce driving and GHG emissions.[17]

Such examples show the potential power of regional-scale planning. If they are directly elected and have sufficient power, regional agencies are best positioned to limit urban sprawl across metropolitan areas through mechanisms such as urban growth boundaries. They can also coordinate public transit systems, establish regional green spaces systems, and promote social equity through programs aimed at affordable housing, equitable schools, livable wages, and tax-base sharing.

Stronger regionalism is usually created by higher-level government. Local governments do not voluntarily give up their power, even though there are important local benefits to regional cooperation. The solution is for national, state, or provincial governments to create a planning framework in which collective goals are set at the highest level—such as clean air, clean water, carbon neutrality, preservation of farmland, compact urban development, equitable distribution of affordable housing, and balanced land use development. Then they delegate power to regional and local governments to implement solutions appropriate to local contexts and make resources available to them.[18] Packages of incentives, mandates, and technical support can ensure that otherwise reluctant local communities meet broader societal goals in terms of land use, resilience planning, and equity.

Adopt Better Forms of Development Regulation

At the local level an important strategy is to develop new types of regulation for land development that are clear to the public, not overly burdensome to landowners, and still protective of social and ecological needs. Form-based codes are one such alternative, developed beginning in the 1990s and promoted by advocates of the New Urbanism. These represent a radically different type of code: conveyed through simple diagrams rather than legalistic text, aimed at form rather than use of buildings, and intended to be intuitively understandable to the public. Usually form-based codes aim at creating walkable, people-oriented urban places in contrast to the suburban, motor vehicle–dependent landscapes produced by many conventional zoning codes.

In practice, some form-based codes have grown complex and unwieldy like traditional zoning codes. The so-called SmartCode is an example, running to seventy-two pages and containing complex tables as well as form types with names like ST-50–26. Still, the form-based approach holds particular promise for downtown and commercial corridor planning, when stakeholders wish to see a particular urban character and wide range of land uses emerge.

Performance standards constitute another code alternative. Here the idea is to specify how a given urban fabric is supposed to perform, for example in terms of keeping stormwater runoff on-site, being zero-net-energy, not creating more than a certain number of vehicle trips per day, or not shading neighboring properties. As long as new development meets these standards, designers are free to configure it in any way they want. Baseline performance standards for cities in the future might call for meeting certain objectives around energy use, climate resilience, and social equity.

A "generative" process in which flexible guidelines, group process, and strong internal values guide individual or collective decision-making is another alternative to traditional land development codes. Past societies in which religious values permeated the culture exemplify this approach. The Zen Buddhist tradition in Japan, for example, with its strong emphasis on simplicity and respect for both people and nature, for many centuries shaped both the design of buildings and the use of land. Islamic culture likewise shaped development of many communities in the Middle East. Within modern society, a generative approach to design has been championed in particular by architect Christopher Alexander, whose book called *A Pattern Language* sought to assemble time-tested principles of good development at many different scales for use in future communities.[19]

Ecologically based management zones represent a final promising direction for land planning. For example, the State of Oregon's Conservation Strategy identifies nine ecoregions across the state, each with particular types of opportunity areas, habitats, and species.[20] Stepping down scales of planning, the Portland Metro regional government has created a Greenspaces Master Plan identifying a metropolitan system of natural areas and greenways, and the City of Portland developed a system of "ezones" that provide additional protection for environmentally important areas, such as along streams.[21]

Promote Compact, Connected, Well-Balanced Land Development

Although the US is the poster child for suburban sprawl, different versions are proceeding rapidly worldwide. Portions of the Italian countryside near Rome are being carved up for upscale estates. Areas north and east of Tokyo are seeing low-density single-family-home development. Lagos is expanding by means of sprawling informal land development on top of creeks and wetlands. China is pioneering high-rise sprawl—constellations of residential skyscrapers outside of major cities without the land use balance, walkable street networks, and transit orientation of older urban areas.

Suburban sprawl's cousin rural sprawl is also rapidly growing. By creating large lots, generally 1–10 acres (0.4–4 hectares), developers try to give residents a sense of living in the countryside close to nature, even though such exurban development rapidly destroys the landscape's value as farmland or natural habitat. Rural sprawl generates very high levels of motor vehicle use and lends itself to class segregation, in different places providing country estates for the rich and refuges for low-income households displaced by the high cost of urban living. As seen in California, development in the wildland-urban interface also contributes to the risk of wildfires. Support for right-wing politicians is typically strongest in these places, perhaps because residents believe they have little in common with those living in more urban locations, prioritize their individualism, and feel that they do not need government because they are living remotely on the land.

The "compact city" is usually the goal of sustainability advocates.[22] The point, though, is for cities to be not just dense but well connected in terms of roads and infrastructure and amenities, contiguous (without development leapfrogging around the countryside), green as well as gray, and with diverse, well-balanced, human-scaled land use. Strong public sector action is usually necessary to ensure these things, plus to constrain sprawl and ensure that builders pay the true costs of urban fringe land development, both infrastructure development and maintenance. Since enormous amounts of money are made subdividing and developing rural land, the public sector must limit the political power of development interests and enact strong policies to protect unbuilt land. This is particularly important to protect food, water, and ecological systems from further development and fragmentation.

Urban growth boundaries (UGBs) or urban service boundaries have been one of the best tools to manage sprawl, since they establish a sharp, clear limit to an urban area, prohibit subdivision of land and/or provision of urban infrastructure beyond a certain point, and are put into place for a relatively long period of time such as twenty years. Portland is one of the few examples of a UGB in the United States, although there are also some around small cities in the northern Bay Area. Canadian cities such as Toronto rely on urban service boundaries, beyond which sewer lines, water lines, and roads won't be expanded, to limit sprawl.

Germany has gone a step further to require extensive coordination between levels of government before land development can take place. Although like the US, Germany has a fairly decentralized governance structure, the national government has established social equity and sustainability goals that local governments must follow. It requires urban regions to prepare strong planning frameworks and engage in consensus-building collaboration. Large areas of land outside cities are designated as off limits to development, a situation that can be changed only with approval of higher levels of government. Landowners in these areas are not entitled to reimbursement for their inability to develop, in contrast to the US legal doctrine of compensating landowners for "takings."[23] The fact that the national government gives local government much of its revenue helps ensure that German cities and towns establish sustainable spatial planning policies, since local governments are not developing land to collect tax ratables.

Land banking is another approach that some European cities use to manage sprawl. If a local or regional government owns much of the land within its jurisdiction, then it can direct urban growth to desired locations such as around transit stations by leasing that land to developers. At the same time it can recapture value from public investment (in this case the rail line) or possibly through mechanisms like tax increment financing, rather than letting those windfalls go to private land speculators. In the US and Canada, the public sector has sometimes purchased suburban land at risk of development for parks and water reservoirs. But generally local governments haven't been major players in land markets. In the future, governments at all scales will probably need to play larger roles in land management in order to help reduce GHG emissions, sequester carbon from the atmosphere, manage region-wide climate adaptation and retreat, and ensure climate justice.

Nonprofit land trust organizations have increasingly stepped into the breach in the US to purchase conservation easements on farmland or open space to preserve them from development. A conservation easement buys development rights from the landowner, to be held in perpetuity by the nonprofit while allowing existing rural land uses to continue. However, this is a less-than-ideal mechanism with which to preserve open space. Wealthy jurisdictions such as Marin County, California, have well-funded land trusts preserving large amounts of open space. Less affluent jurisdictions have no such resources. This equity disparity is another reason why growth management planning is necessary at higher levels of government, so that all cities and towns have equal ability to preserve unbuilt land and achieve the benefits that this provides.

Along with steps to limit outward sprawl around cities, policies to promote infill development are essential. Otherwise new housing and commercial development have nowhere to go. In the US, policies that redirect urban growth to infill locations are often known as smart growth and are justified on grounds of reducing long-term public costs of providing infrastructure. The state of Maryland was a smart growth leader beginning in 1997, when it restricted state transportation funding to locally designated Priority Funding Areas, while also supplying state funding for local preservation of Rural Legacy Areas. Maryland's smart growth efforts have had ups and downs over the years with political changes, but the idea of using state or national funding to leverage better local planning and development is a promising strategy for many places worldwide in the future.[24]

Local zoning codes currently restrict much urban land in the US to detached, one-or-two-story, single-family home development on large lots. Upwards of half the land in American cities is zoned this way. Cities are just beginning to overhaul these restrictive zoning frameworks to allow additional units on each parcel of land. Minneapolis, for example, has changed its single-family zoning to allow three units on lots that formerly were restricted to just one. The State of Oregon approved a bill in 2019 allowing the development of duplexes and townhomes in areas zoned single-family. Upzoning land in this way, especially near transit stations and neighborhood centers, will be essential to create an overall metropolitan land use pattern that works well for sustainability and promotes equity. Upzoning efforts will need to be mindful of how increased density will change pressures on schools and other community amenities.

Encourage Finer-Grain Land Development

One basic problem of twentieth-century development was its seemingly ever-increasing scale. Commercial, office, and industrial buildings as well as residential subdivisions all grew in size, with 220,000-square-foot Walmart Supercenters perhaps the ultimate example in terms of commercial development. Instead of homes being built one by one, production home-building companies built them by the thousands using assembly-line procedures. Street widths expanded as well, with many suburbs now oriented around six- or eight-lane arterials. In these ways, communities remade themselves according to the scales best fitted to motor vehicles and global capitalism, not to humans.

The New Urbanism movement that began in the 1990s emphasized a return to pedestrian-oriented, human-scaled communities with short blocks, neighborhood centers, and mixed housing types and land uses. However, it was not totally able to escape the bigness problem itself. Since New Urbanist communities were typically created by large development companies operating at their accustomed scale on enormous sites, many suffered from a suburban monotony similar to conventional suburbs. They also tended to reproduce existing income-segregated communities. And although they were pedestrian oriented, many New

Urbanist communities were in suburban locations accessible only via large highways, so people had to drive to get to them.

One strategy to respond to the problem of excessive scale is for local governments to work with citizens to hear their concerns and then help them reimagine more sustainable and equitable urban forms. Local government can require mixed types of buildings and uses as well as affordable housing within new neighborhoods and sharply limit the number of buildings that a single developer can build. This would ensure a higher degree of variation. Tiny homes and accessory dwelling units (ADUs) on single-family-home lots could be encouraged as well. Plans for new neighborhoods should also create a highly connected street network with small blocks, small but varied lot sizes, a well-connected green spaces system, and highly efficient neighborhood-scale energy systems. The end result would be a smaller-scale, more varied, and more equitable pattern of development.

Preserve Ecological Function during Land Development

Cities, towns, suburbs, and exurbs have been built for many centuries with little attention to the landscapes and ecosystems that they exist within. Developers have flattened hills, filled wetlands, put waterways into pipes or culverts, bulldozed wildlife habitat, and paved over the soil. In the long run this mode of development doesn't work well either for people or for ecosystems. One main strategy for the future is thus to build more in balance with nature, integrating patches and corridors of wildlife habitat into cities. The discipline of landscape ecology that arose in the late twentieth century can serve as a guide.[25] As part of this strategy cities can protect and where necessary restore hydrological systems—streams, rivers, floodplains, and wetlands. Ecological function can also be integrated into buildings, for example through green roofs, walls, courtyards, and interior spaces. These ecological planning tools have the benefit of helping communities adapt to global warming and associated climate risks.

Planning and design processes as well as code changes can help this urban-ecological integration come about. For example, the public sector can require that development be set back from streams, rivers, wetlands, and floodplains and that it handle stormwater runoff on-site. It can require new subdivisions to preserve wildlife corridors and significant patches of habitat. And it can provide incentives or grant funding for restoration of both waterways and habitat. California, for example, has had an Urban Streams Restoration Program since 1985 that has provided 270 grants of up to $1 million to local communities, allowing them to dig buried streams out of pipes, recreate stream channels, improve riparian vegetation and habitat, and involve local residents as stewards of their landscapes.[26] Chapter 8 will further explore the topic of greening cities.

Set Maximum Limits to Land Ownership

Highly unequal distribution of land is a sustainability problem in many countries. In the US, eight families each own more than one million acres apiece, an area larger than the state of Rhode Island.[27] Peel Holdings, a secretive company controlled by the billionaire John Whittaker, owns large swaths of the British cities of Manchester and Liverpool, plus airports and rural areas where fracking for oil and gas is carried out.[28] In the Middle East, the emir of Kuwait owns approximately seven million acres, basically the entire country except for a large American military base.[29] In Guatemala, the largest 2.5 percent of farms take up almost two-thirds of available farmland, while 90 percent of farms are on only one-sixth of the land,[30] often in mountainous areas with steep slopes and poor soils. These conditions lead to poverty, deforestation, and a flow of landless migrants from the countryside into cities and abroad into other countries.

Land redistribution is the most straightforward strategy for excessive concentration of ownership. In the 1950s, Guatemalan president Jacobo Arbenz and Indonesian president Sukarno attempted to redistribute lands within their countries. However, both were overthrown militarily by wealthy factions with active support from the American CIA. Cuba, Nicaragua, and Zimbabwe had greater success with land redistribution in the 1960s (for Cuba) and the 1980s (for the other two). However, they also suffered retribution from the US and other nations who did not like this challenge to capitalist land ownership. In the 1960s Egypt also issued laws that limited large-scale ownership of cultivated lands and redistributed lands from the rich to the farmers who were working them.

Many of the nations of eastern Europe, including Lithuania, Romania, Yugoslavia, Czechoslovakia, Poland, Bulgaria, Finland, and Hungary, engaged in land redistribution programs after World War I.[31] Finland, for example, made landholdings larger than 200 hectares (494 acres) property of the state. Poland reallocated landholdings larger than

180 hectares (445 acres) to landless peasants. Germany also redistributed substantial amounts of land.

Within wealthy countries currently, milder forms of redistribution occur through urban redevelopment processes under which local agencies expropriate blighted or underused properties with compensation to owners and redistribute them to new owners who will presumably use them better or for public purposes, often adding infrastructure, amenities, and affordable housing in the process. Municipal land banks can also make land and housing available to the poor. Community benefits agreements are another type of redistribution in which developers provide land and amenities to communities in return for local organizations supporting their projects. These arrangements acknowledge the impact that development can have on communities and seek to redress them with community benefits.

A further strategy to ensure broad ownership of land is to limit absentee ownership. New Zealand has restricted international ownership of domestic residential property in order to reduce price inflation. Victoria, British Columbia, taxes absentee owners an extra 2 percent of property value to discourage such ownership. In rural jurisdictions, limiting absentee ownership of farmland would be an excellent way to ensure that young would-be farmers could access land. Such a concept builds on the ancient notion of usufruct rights—that those who use land productively without damaging it should have rights to it.

Recapture Speculative Gains in Land Value

On the theory that society as a whole creates value for particular pieces of land, nineteenth-century philosopher Henry George proposed that the economic value derived from land should belong to the whole of society rather to individuals and that land should be common property. George's preferred mechanism to bring this about was a "single tax" on land value (rather than on buildings constructed on the land).[32] Such a tax would discourage landowners from holding urban land vacant and would also discourage vast private rural landholdings if the owners were not using the land. George's position was closely echoed nearly a century later by the United Nations 1976 Habitat 1 Declaration, in which nations agreed that "the increase in the value of land as a result of public decision and investment should be recaptured for the benefit of society as a whole."[33] This increase is often known as the "unearned increment" of land value.

George's "single tax" has been put into practice only rarely. After gaining independence from the Soviet Union, Estonia adopted a land tax in 1993 to discourage speculative holding of property.[34] Cities such as Porto Alegre, Belo Horizonte, Bogotá, Mexico City, Johannesburg, Cape Town, and Pretoria have taxed vacant land at significantly higher rates than developed property as a way to reduce land speculation. Many jurisdictions in metropolitan Manila levy an idle land surcharge of 5 percent.[35] At the national level, Denmark, Australia, New Zealand, and several African countries use land taxes to some extent.[36] New Zealand relies on property taxes to fund 60 percent of local services,[37] thus penalizing large landowners.

Expand Collective Ownership and Management of Land

It has been taken for granted in international development circles for most of the last century that "regularization of land tenure," that is, the establishment of clear individual ownership of land, is necessary for progress. Accordingly, common-property traditions are being wiped out worldwide, with newly privatized land entering land markets and often exploited within capitalist economies. But what if this is not the case?

One radical strategy for sustainable development is to reclaim and reinvigorate the commons. Here the work of Elinor Ostrom (1933–2012) is crucial. More than anyone else, she has shown how collective management of common-pool resources can potentially surmount the "tragedy of the commons" lamented in 1968 by Garrett Hardin,[38] in which individuals acting rationally according to their own self-interests destroy resources necessary for collective survival. Winner of the 2009 Nobel Prize in Economics (and the only woman ever to win the Nobel Prize in Economics at that point, which says something about the economics profession), Ostrom showed how traditional and modern peoples have in fact managed common-pool resources in ways that conserve those resources and ensure long-term prosperity. She identified a number of principles for stable management of common resources, including clear definitions of the resource, shared decision-making, effective monitoring, sanctions for violations, and multiple scales of management with a degree of self-determination at each level.[39]

In terms of land, the common-pool management approach could support various forms of collective ownership. A given group of local residents, community, tribe, or users (such as foresters or fishermen) might manage an area of land, including its physical landscape, hydrology, and ecological community, with a common title to the "property" plus guidelines or requirements for sustainable management.

As mentioned previously, community land trusts are a modern type of organization to exercise such collective management. These non-profit groups usually have a board composed largely of local community members and a set of goals based on local needs. They may hold and develop land, for example to provide affordable housing and jobs for local residents. They can work in conjunction with municipal land banks to make sure that vacant land is used for community benefits. Examples in the US include the Dudley Street Neighborhood Initiative in Boston and the Sawmill Community Land Trust in Albuquerque. Other land trusts focus on preserving farmland and open space and often hold conservation easements on such land rather than title. A preeminent example is the Marin Agricultural Land Trust, which has managed to preserve eighty-five family farms totaling fifty-three thousand acres from suburban development in Marin County, a rugged and scenic coastal jurisdiction immediately north of San Francisco.[40]

Societies could also establish new types of commonly held property. The ejido system in Mexico is an excellent example. During and after the Mexican Revolution of 1910–20, landless peasants demanded access to land that was concentrated in the hands of wealthy families. The administration of President Lázaro Cárdenas (1934–40) set up a system in which groups of peasants were given control over land owned by the wealthy. Title was held by the Ministry of Agrarian Reform, but these ejidos were allowed to live on and collectively manage the land provided they used it productively. By the 1990s some twenty-eight thousand ejidos with more than three million members controlled more than 230 million acres of land—almost one-third of the country. Although recent neoliberal Mexican administrations have sought to privatize ejido land, the system to a large extent remains in place.[41]

CONCLUSION

Managing land sustainably has many challenges. It means first and foremost rethinking public attitudes toward land ownership—for example, to move away from the notion that one person should be able to do what they like with a piece of the Earth. It means active efforts to manage urban growth, preserve farmland and open space, and promote revitalization of existing urbanized areas that addresses racial justice and does not promote gentrification. It means rethinking zoning, subdivision, and building codes, as well as prevailing patterns and scales of development, particularly in the face of climate risks. It may mean setting limits on private ownership and establishing new institutions, as when Mexico created the ejido system in the 1930s, or when activists began creating community land trusts in the US during the late twenti-eth century. And it means working tirelessly to ensure that everyone has access to decent housing and green spaces.

As usual, the best strategies for any given place will depend on the context and community input. But everywhere improvements to how we use land are possible. Sustainable land use is key to our long-term survival on the planet, and radical changes in land management and ecological planning will be essential to create more sustainable cities and societies.

8

How Do We Design Greener Cities?

In Hammarby Sjöstad, a dense urban district of four- to six-story apartment buildings built in the 1990s two miles south of the center of Stockholm, all residents can get to green spaces simply by walking out the door of their building. Paths and canals lined with vegetation cut through the neighborhood. Vegetated courtyards fill the center of most blocks. Children play along trickling streamlets. "Ecoducts" take pedestrians and a broad swath of nature over an expressway. Motor vehicles and urban noise are kept to a minimum. Most dramatically, shoreline walkways give residents views across the water to the islands to the north (figure 10).

Every city or town could be this way if we decided to make green cities a priority—with nature woven into the fabric of daily life. Imagine communities dense enough to have urban amenities, such as close-by stores, cafes, restaurants, performance spaces, and public transit, but also quiet, green, and safe. Imagine neighborhoods with vibrant urban plazas and boulevards but also community gardens, greenway networks, green roofs, and wetlands. Imagine being able to encounter nature easily outside one's door.

Biologist E. O. Wilson used the term *biophilia* in 1984 to refer to "the urge to affiliate with other forms of life,"[1] and Timothy Beatley applied the concept to cities in his 2010 book *Biophilic Cities*.[2] Beatley envisions urban communities where children can learn from natural systems, where natural forms and images are integrated into buildings, and where residents benefit from restorative time in natural spaces.

FIGURE 10. For the Hammarby Sjöstad neighborhood in Stockholm, the city repurposed old industrial land as a dense but very green community knit together by pedestrian walkways. (S. Wheeler.)

The grayness of our current cities and towns has had many negative effects, ranging from the creation of urban heat islands and degradation of watersheds to stresses on our mental and physical health. Generations of city dwellers in many countries have moved to greener suburbs, contributing to suburban sprawl, regional inequity, and its resulting negative externalities. This movement frequently has had racial and class dimensions as well. In the US, the growth of metropolitan regions and accompanying patterns of greening are the product of structural racism and deliberate discriminatory practices such as redlining, restrictive racial covenants, and the denial of FHA loans to Black Americans. As the middle and upper classes moved away from cities toward greener suburbs, they took tax base and economic investment with them, and central cities became disinvested, often losing tree cover and park maintenance funds. Though many have experienced reinvestment in recent years, the need for urban greening without gentrification and displacement still remains.

Urban greening is in large part a question of urban design—the creative arrangement of spatial forms.[3] It takes careful thought and motivation to bring nature into cities. Spaces must be reserved for vegetation, even if these are rooftops and walls. Green spaces must be cared for and need to support surrounding communities. Water drainage systems must be thought through so that they incorporate natural principles. Habitat value and other ecological benefits must be maximized. Equity needs to be a guiding value. Design is not a one-time action but an ongoing process of engaging with diverse groups of residents to improve and care for the environment, taking stewardship over the places in which we live.

Urban greening is also a question of social policy and values. It requires spending money on recreating natural systems and forgoing the revenue from building as densely as possible. It requires—given a commitment to social equity—providing parks and green buildings for people who cannot pay for them. It also requires preventing green space investment from leading to gentrification and displacement, for example, by putting affordable housing in place before we focus on greening.

Green spaces provide important ecological, social, economic, health, and psychological benefits, a new calculus that creates greater value in the long run than the purely economic "highest and best use" of land that has prevailed in the past. Imagine how hot and oppressive New York City would be without Central Park! Imagine London without being able to walk along the Thames, or Paris without being able to walk along the Seine! Greening is a main way to make urban places highly livable. The result will be well worth it: easy access to nature for everyone within human communities, ecological benefits for flora and fauna, cleaner air and water, opportunities for children and the elderly to be in healthy outdoor spaces, and the simple delight of quiet, green oases amid the excitement and vitality of diverse human communities.

A GROWING IMBALANCE OF CITY AND COUNTRY

The balance between city and country—gray and green—has been an ongoing debate since the early nineteenth century. Until then most human settlements were small and compact, and there was little need for large-scale parks and green spaces planning. Urban edges were sharp, and suburbs as we know them now did not exist. Most residents could walk from their homes into the countryside quite easily. Farms, fields, and forests were within eyesight. There were no motors to fill the air with noise. The main sounds were made by people, animals, and occasionally the wheels of carts or carriages. When night fell the world grew dark. Nature was much closer at hand.

As the nineteenth century went on, human communities became increasingly divorced from nature. Virtually unregulated industrialization and enormously overcrowded worker housing made cities grim places. Unfiltered coal smoke dirtied the air. Effluent from factories, slaughterhouses, tanneries, and sewage polluted waterways. Disease was rampant. Cities, arguably, were the first large-scale "sacrifice zones" in terms of industrial development and environmental justice.

Such conditions led to a debate about how to balance cities with nature that continues to this day. Specific projects tried to pioneer a new balance. Nineteenth-century North American neighborhoods such as Riverside in Chicago and Rosedale in Toronto began to create greener urban districts for the wealthy, using picturesque design elements such as gently curving streets, larger lawns and planting strips, and small parks. In the 1859 Eixample district of Barcelona, Spanish engineer Ildefons Cerdà allocated half of each block to green space and truncated the corners of blocks to create small plazas at each intersection. The result was a pioneering attempt to create a green city. However, the lure of profit was too great, and after the neighborhood was laid out, land-owners filled the entire blocks with buildings. Still, the Eixample district remains a lovely and livable neighborhood of Barcelona to this day. Barcelona is now exploring the related idea of retrofitting neighborhoods as car-free "superblocks" to promote urban greening, reduce motor vehicle use, and clean the air.[4]

A better-known and even more comprehensive vision of how city and country might be balanced was Ebenezer Howard's 1898 Garden City proposal. This called for a regional ring of garden cities, connected by rail, surrounding an older central city. Each of the new towns would have a walkable urban core, concentric boulevards, and a sharp edge against the surrounding countryside. Howard's vision became one of the most influential in urban planning and influenced later generations of urban planners and designers who sought to decentralize cities and add vegetation. Unfortunately in practice many developers simply built low-density residential neighborhoods without the walkable centers, mixture of uses, and countryside preservation that Howard had advocated.

The Parks Movement of the late nineteenth and early twentieth centuries led to new green spaces within cities, the most famous of which are Central Park in New York City, Prospect Park in Brooklyn, and the Fens in Boston. These were a major step toward greening cities.

However, typically such efforts created just one or two large parks within a given community and presented a romantic facsimile of nature rather than functional habitats. As historian Galen Cranz documents, early North American park designs borrowed heavily from the country estate designs favored by English gentry, with large and sweeping expanses of mown lawn beneath widely spaced shade trees.[5] Central Park is in many ways a romantic redesign of both natural and human elements previously on the site. Continental European urban parks often took on even less naturalistic forms with extensive gravel walkways, manicured hedges, reflecting pools, and ornamental fountains. Lacking understory and a range of plant species, parks within the city often had little ecological value.

Park development in many parts of the world also reproduced the colonial, racist, and sexist view that white men should make the final decisions about the use of land and that the wealthy had a right to access natural spaces that others could not. For Central Park, developers removed several villages of free blacks and Irish immigrants, filled swamps and lakes, and resculpted the land with massive amounts of gunpowder. The five-acre community of Seneca Village, founded by free blacks in 1825, was taken by eminent domain and razed so that the park could be built.[6] Only recently have planners and historians started to reckon with the troubled history of park planning.

As formal urban planning and landscape architecture professions emerged during the first half of the twentieth century, they helped develop new codes and standards that specified ways to green human communities. Unfortunately, the main way that they did this was to require low development densities, large minimum lot sizes, and large setbacks between buildings and property lines. Ample private green space existed within the resulting suburbs, but the resulting lawn-scape again had little habitat value. In fact, suburban sprawl led to habitat fragmentation and both air and water pollution. Moreover, much of the landscape was now covered by asphalt for streets, driveways, and parking, which led to rapid stormwater runoff and degradation of hydrological systems. Although early twentieth-century regionalists such as Patrick Geddes and Lewis Mumford had called for decentralization of the metropolitan region in order to create a better balance of city and nature, and are sometimes known as ecological regionalists, the result was anything but ecological.

The twentieth-century urban development codes that remain on the books often make both urban greening and compact urban development difficult. For example, it can be hard for a developer to cluster housing units on a portion of a site so as to preserve the rest as open space if those denser units violate height or minimum lot size requirements. A landowner wishing to cover much of a parcel with a large urban building but to provide rooftop green space often has difficulty getting the roof to count toward open space requirements for new development. In such ways the codes developed in order to decentralize and green the overcrowded nineteenth-century city often hinder today's creative efforts to improve the ecological function of cities and promote affordable housing.

The effort to balance city and country took a new turn two-thirds of the way through the twentieth century with the rise of the modern environmental movement. This new public ethic of environmental protection, translated through organizing efforts into political and institutional strength, began the process of cleaning up urban air, water, and soils as well as restoring ecosystems previously damaged by development. Although in the 1960s and '70s some people left urban areas to move "back to the land" and some left for reasons related to race or class, others stayed in cities and suburbs with a growing consciousness about how their daily lives intersected with the natural world. Simultaneously, many urban residents became concerned with personal health and exercise. Rachel Carson's *Silent Spring* in 1961 sounded a warning about the dangers of toxic chemicals in the environment, and rapidly worsening mid-twentieth-century air quality in cities such as Los Angeles and London was a visual symbol of growing pollution. (Developing-world cities such as Beijing and Delhi are experiencing this situation in the twenty-first century.) These changes fueled efforts to create greener urban environments as well as a growing awareness of environmental injustice.

One main front in late twentieth-century urban environmentalism was the struggle against suburban sprawl. In 1961 Jean Gottmann noted that on the US East Coast it was no longer possible to identify where cities stopped and countryside began—urban and suburban areas had merged into one giant "megalopolis."[7] Other megalopolises began to cover Southern California, parts of Northern California, areas around the Great Lakes, parts of Texas, Germany's Ruhr Valley, the Nagoya-Osaka-Kyoto-Kobe corridor in Japan, China's Pearl River Delta area, and the Rio de Janeiro-Sao Paulo region within Brazil. Truly rural or wild lands became difficult to find near urban areas. Instead, cities were now surrounded by a wildland-urban interface that was really neither.

Many environmentalists' emphasis in the late twentieth century shifted from preserving unbuilt land to restoring previously damaged ecosystems. The new field of landscape ecology informed such efforts. This discipline developed a language for analyzing ecosystem spatial patterns and dynamics. Landscape ecologists studied "patches" of habitat, "edge" and "interior" environments catering to different species, "corridors" necessary for species to travel from one patch to another, and "matrices" that combined all these elements. Implications for urban green spaces planning were clear. Previously isolated urban parks or remnant undeveloped areas had to be connected with one another as well as with riparian corridors (along waterways). Planners and activists now had a more sophisticated understanding of how to create a regional green spaces system with high habitat value. More recently a "rewilding" movement has sprung up that takes ecological restoration to a new level, calling for returning previously developed land to nature.[8]

Urban environmentalism took on a new dimension in the late twentieth century when the environmental justice movement called attention to racial and class disparities in environmental impacts and access to green space. Low-income, Black, Latino, and Native American communities in particular had long suffered from disproportionate exposure to toxic waste dumps, landfills, polluting industries, and air pollution. Then and now they face higher rates of cancer, diabetes, asthma, and obesity and have shorter life expectancies. Disparities in local government finances mean that disadvantaged neighborhoods often have far less access to parks and inferior maintenance of green spaces. Public interest organizations such as the Trust for Public Land documented that the poorest neighborhoods of color in New York (in the Bronx and Queens) had fewer decent parks than affluent neighborhoods, and that residents in relatively poor cities such as Mesa, Arizona, and Oklahoma City were much less likely to have parks within walking distance of homes than those living in wealthier cities.[9]

Further urban greening themes emerged at the turn of the millennium. Beatley argued for biophilic cities that would connect people with nature on a daily, personal level. Examples include Austin, Texas, where residents gather each evening to watch bats stream out from their roosting places underneath one of the city's bridges, and San Diego, California, where residents have taken stewardship over the steep canyons that cut through many of the city's neighborhoods. Richard Louv emphasized the necessity of connecting children to nature in his book *Last Child in the Woods*. Urban agriculture gained popularity. Movements such as foraging—seeking wild plants, mushrooms, and other materials to eat, even in urban locations—likewise illustrate a biophilic approach to urban life. Apps such as iNaturalist and iSeek have helped citizens identify species in their communities and act as citizen scientists and ecological advocates. European eco-districts such as Western Harbor in Malmo and Hammarby in Stockholm weave nature through newly constructed urban neighborhoods.

Many cities undertook tree planting and on-site drainage programs for climate adaptation purposes. Public health authorities in many countries embraced "active living by design" in order to counter the obesity epidemic sweeping industrial societies worldwide—a movement that assumes that urban residents have walkable streets, trails, greenways, and parks at hand for purposes of physical activity. The Covid-19 pandemic showed the importance of outdoor spaces for people sheltering in place, and the inequitable distribution of these across communities. Providing these amenities in low-income communities of color—and at the same time addressing the social justice problems those communities suffer from—emerged as a clear challenge for the future, particularly given the disproportionate impact of climate.

STRATEGIES FOR GREENING CITIES

Today a broad movement to green cities is under way. Greening, of course, needs to be done in conjunction with other policies to reduce economic inequalities, address racial injustice, improve housing, provide transportation alternatives, and meet other sustainability and equity needs. It becomes part of the new ethic of sustainable development in which development projects of all sorts simultaneously heal ecosystems and communities—in the process generating healthier social ecologies.

Several kinds of strategies can help take the integration of cities and nature to this next level (table 8).

Plan for Regional Green Space Networks

A starting point for each urban region is to develop a strong green spaces plan and identify sufficient resources to implement it. This is best done at a metropolitan scale, since watershed features and other green spaces typically overlap the boundaries of individual cities and towns. Planners can analyze ecosystems and use geographic information systems (GIS) to map priority areas for conservation or restoration as well as neighborhoods that are underserved by parks and green spaces.

TABLE 8 STRATEGIES FOR URBAN GREENING

Strategy	Description
Plan for regional green space networks.	Develop regional and/or municipal plans identifying green space systems and policies and programs for enhancing and networking these, with special attention to social equity dimensions of access to green spaces.
Restore ecosystems.	Adopt programs to remediate pollution, revegetate landscapes, improve hydrologic function, and restore habitat, engaging community organizations for ongoing stewardship.
Encourage eco-districts.	Jump-start programs to improve sustainability at the neighborhood scale, for example with microgrids of renewable energy, district-scale heating and cooling systems, on-site drainage, and urban agriculture.
Expand green infrastructure.	Pursue citywide strategies to increase tree canopy and vegetated cover, especially in disadvantaged neighborhoods.
Create green streets.	Add vegetation to planting strips, medians, or curb bulb-outs along streets so as to handle runoff on-site, slow traffic, and improve pedestrian comfort and experience.
Encourage green roofs and walls.	Require or incentivize vegetated rooftops and walls to cool the city, slow runoff, add habitat, and create restorative spaces.
Reduce hardscape.	Replace asphalt and concrete with permeable surfaces to reduce runoff and cool cities.
Promote climate-appropriate landscapes.	Require or encourage low-water vegetation around buildings for aesthetic, recreational, dietary, and habitat needs.
Expand urban agriculture.	Provide physical space and incentives for community gardens, allotment gardens, market gardens, greenhouse agriculture, beekeeping, and urban animal husbandry.
Reduce noise and light pollution.	Require quieter motor vehicles and well-shielded outdoor lights; reduce roads and exterior lighting.
Avoid green gentrification.	Proactively develop programs to add affordable housing and reduce displacement; engage neighborhood residents to ensure that needs of all are being met.

Creating a green infrastructure equity index can help with this prioritization.[10] Plans can then coordinate open space preservation, habitat restoration, flood prevention, trail systems, urban agriculture, and social equity considerations such as affordable housing and park and playground access. An example of such planning at the regional scale is provided by the Metro Council in Portland, Oregon, which has developed a regional green spaces plan, manages seventeen thousand acres of parkland, has required local governments to adopt environmentally oriented land use planning policies, and if necessary can condition infrastructure funding on local compliance. Portland has also made green space equity a key requirement.

In the absence of strong individual agencies to plan and fund green space networks, advocates can craft planning frameworks between multiple players, often putting the pieces together over many years. For example, environmentalists and elected leaders got the State of Washington to pass a 1990 Growth Management Act requiring the state's fast-growing cities to address thirteen planning goals, including "open space and recreation" and "environmental protection," within their Comprehensive Plans. Local officials in the city of Seattle then approved Comprehensive Plans in 1994 and 2016 with ambitious Parks and Open Space elements. But the city lacked funding to acquire new green space land, so environmental leaders asked voters in 2014 to approve a new Seattle Park District and increase their property taxes to supply approximately $50 million a year. Voters did this, and then the city drafted a more detailed 2017 Parks and Open Space Plan to define acquisition and investment priorities. Planners also updated the city's Capital Improvement Program to include green spaces and conducted a Recreation Demand Study.[11] These initiatives together helped coordinate planning for a 6,414-acre system of 485 parks, recreational facilities, trails, and natural areas.

Many other types of agencies and organizations can develop urban greening plans, keeping in mind that access to green spaces should be a right for everyone and that existing inequities in this regard need to be addressed. Resource conservation districts are nonprofit agencies established under state law in the US to improve land and water conservation practices, working with private landowners. Nearly three thousand exist, almost one in every county, and they have been a time-tested way to improve management of soils and waterways.[12] Land trusts are nonprofits set up to protect open space and habitat in rural areas, although in cities they can also have a mission of providing affordable housing and community facilities. They often purchase land or development easements in order to protect ecosystems and manage the land to restore ecological value. Various public-private partnerships have created and implemented Habitat Conservation Plans required under the federal Endangered Species Act.[13] The Santa Monica Mountains Conservancy, for example, has preserved seventy-five thousand acres of undeveloped land near Los Angeles in this way. Established by the California Legislature in 1980, its board includes members appointed by the mayor of Los Angeles, the state's governor, the legislature, multiple local counties, and local park agencies.[14] These are just a few examples of the many organizations that can take action to promote urban greening.

Restore Ecosystems

The move toward ecological restoration has coincided with the deindustrialization of large areas within North American and European cities, meaning that large amounts of brownfield land have become available for reclamation. Early projects included restoration of green spaces within Germany's Ruhr Valley as well as riverfront parks within Pittsburgh—not coincidentally both former centers of the steel industry.[15] Often designers kept remnants of the former mills, factories, and industrial infrastructure as a way of honoring an area's history.[16] Any abandoned industrial site is now a leading opportunity for urban greening. Cities can actively identify these and initiate planning processes.

Shorelines represent another major opportunity for restoration and recreation. In the late twentieth century, London put walkways along the Thames and built a pedestrian bridge arching over the river to the new Tate Modern museum (itself housed in a reclaimed power plant). The San Francisco Bay Area launched planning for a five-hundred-mile Bay Trail around the entire bay. Toronto developed restoration plans for its two main riparian corridors, the Humber and Don Rivers, and turned an enormous landfill extending into Lake Ontario into Tommy Thompson Park. Such examples show how cities can provide residents renewed connections to local bodies of water.

Landfills represent a third main opportunity area worldwide. Although once on the urban edge, many old dump sites are now near the center of metropolitan areas and thus can provide millions of people with green space. In the 2000s, Cairo converted a five-hundred-year-old waste dump in the middle of its old city into El Azhar Park, which includes social services in addition to green spaces.[17] Tainan, Taiwan, used volunteer labor to turn a landfill into Barclay Memorial Park. New York converted its own gargantuan landfill, the world's largest, into the 2,200-acre Fresh Kills Park on Staten Island.

Restoration efforts have been informed by landscape ecology and movements for use of native species and involvement of local communities as green space stewards. Such synergies have created a fifth main era

of park design according to Galen Cranz and Michael Boland: the ecological or sustainable park.[18]

Encourage Eco-Districts

Eco-district planning seeks to maximize landscape and environmental
performance at a district scale within new neighborhoods. These communities have been built primarily in northern Europe, where some of
the best-known examples are Hammarby, Western Harbor in Malmo,
and Vauban in Freiburg. Typically eco-districts are quite dense and urban
in flavor, composed of three- to six-story apartment buildings with ample
green spaces around them. They allow for district-scale heating, cooling,
and waste systems plus well-thought-out land use, transportation, and
green spaces systems. Hammarby, for example, has its own light rail
extension, district heating plant, methane generation facility (where
cooking gas is collected from the sewage treatment process), and pneu-
matic tube waste and recycling collection system.

Eco-districts are particularly well suited to brownfields redevelopment, since old industrial or commercial landscapes (when environmentally remediated) offer wonderful locations for new neighborhoods.
Former railyards, failed malls, old military bases, and unneeded port
areas can be redeveloped in this way. Because of their large scale, eco-
districts typically require proactive planning to develop district-level
infrastructure, urban design, and green spaces networks. One main
challenge is for local governments to ensure that such redevelopment
includes affordable housing and is conducted by creative small-scale
local developers who understand and support local culture and assets,
rather than large multinational companies who will likely produce a
more bland, generic, homogenous, high-end product.

Expand Green Infrastructure

Strategic creation of "green infrastructure" is a related approach to
urban greening. Green infrastructure uses ecological systems to provide
services to humans such as flood protection, stormwater management,
improved air quality, water purification, groundwater recharge, food
production, sewage treatment, green space, and recreation. For example, the Dutch "Room for the River Program," begun in 2006, removed
dykes and concrete channels along four of the country's rivers to restore

floodplains and create more natural river flow.[19] This €2.2 billion initiative included thirty-four separate projects in urban areas such as Nijmegen, Arnhem, and Utrecht as well as within rural watersheds. In addition to flood protection, benefits included improved habitat value and riverfront parks.[20] Project organizers had to coordinate involvement by stakeholders at many levels, in this case local, regional, national, and European Union scales.[21]

"Blue" infrastructure focuses on use of water and waterways.[22] Water in the form of fountains, reflecting pools, canals, rivers, streams, and runnels has been a centerpiece of cities for millennia. Water attracts people, especially children. It creates sound, moves, reflects light in interesting ways, and cools the environment around it. On warm days people love to dangle their feet in water or immerse themselves in it. Making water visible and accessible to urban residents complements urban greening. Indeed, "blue-green cities" have become a new ecological planning ideal.

Keeping storm drainage aboveground through use of swales, runnels, and small retention ponds is one way to make water systems visible within cities. Streams can be dug out of pipes and brought aboveground as central features of urban spaces. Stormwater runoff and gray water from buildings can be channeled into holding tanks for later use as irrigation. Pathways can be created along bodies of water, and shorelines reclaimed to become part of public trail systems. Fountains and "splash ponds" can be designed so that people and their children can get wet and have fun on hot days. Excellent examples can be found at Philadelphia's City Hall and Boston Common. Freiburg, Germany, benefits from its system of channels at the edge of streets that formerly removed sewage and other waste from the old medieval city. Now these are magnets for children on warm days.

Of course, many of these strategies are difficult for cities in arid or semiarid climates. Use of water there must be balanced with the need for conservation. But some water features may still be desirable in highly visible or heavily trafficked locations. Moving toward sustainability is a balancing act, and sometimes it is worth using a bit of a scarce resource (in this case water) to create an important public amenity or civic gathering place. Water features can also recycle the water they use or use gray water rather than potable water.

In the era of climate change, radical programs to shade cities are needed in many parts of the world, as discussed in chapter 2. Although many of the world's urban regions have long had trees in them, "urban forestry" implies that these are numerous and continuous enough to yield real benefits in terms of cooling and cleansing the air, reducing noise, providing wildlife habitat, and helping keep stormwater on-site. Creating urban forests is not easy, however. Tree-planting programs need to be designed carefully, for example choosing resilient species that can handle small planting spaces, compacted soil, little water, and multiple forms of pollution. These programs need to address barriers to adoption such as liabilities associated with sidewalks, trees, water and sewer pipes, and gas lines. Arid and semiarid cities will need species that provide shade but use little water. Large trees may not in fact be desirable in cold, high-latitude cities that need as much sun and warmth as they can get, or in spots elsewhere where building owners want to maximize solar access for photovoltaics. In a warming world, trees must also be planted that will withstand the climate of the future. Despite being an iconic native tree, redwoods, for example, may no longer be viable in most of Northern California.

Especially in dry climates, multistoried, low-water plant communities can create shade and improve wildlife habitat while using far less water than traditional turf-and-tree designs. Maintenance needs for such landscapes may be higher than for the conventional type of park, but on the plus side less fertilizer and herbicide is needed, and fossil fuels for mowing are not required. Good design is important to balance vegetation with needs for open lines of sight through urban parks, so as to avoid creating hidden places that could be safety problems.

Create Green Streets

Streetscapes are often responsible for cities' "gray" image. Yet streets are relatively easy for municipal governments to green. The city owns street rights-of-way and must regularly repave or otherwise maintain roads. Each repaving is an opportunity to rethink a street's design. Higher-level governments often provide large amounts of funding for roadwork from gas taxes, vehicle registration fees, or other sources. Growing movements for bicycle and pedestrian planning as well as public experience of "open streets" during the Covid-19 pandemic support street redesign and reprioritization.

Green street design seeks to convert unneeded portions of the street right-of-way into vegetated planting strips or swales. Since in many cases streets are wider than necessary for the amount of traffic they carry, road diets have become popular in many cities. A typical road diet converts a four-lane arterial street to three-lanes with two travel lanes plus a central turn lane. This allows the extra space to be used for bike lanes, sidewalks, and/or greening. Street designers can also add bulb-outs (curb extensions) where pedestrians cross at intersections; these both enhance pedestrian safety and create planting opportunities. Linear swales with riparian vegetation (plant species typically found along waterways) can often be added beside sidewalks to handle stormwater runoff on-site. Designers can improve street tree health by expanding permeable areas of soil and artificial surfaces around them. These many small design interventions can work together to create a greener, healthier, safer street environment.

A leading example of a citywide policy for green streets is Portland, Oregon's, 2017 Livable Streets Strategy. This plan "encourages people to get creative and re-imagine their streets, parking spaces, plazas, and alleys as places to enjoy and engage the surrounding community."[23] A related Portland in the Streets program encourages block parties, street paintings, play streets, street fairs, street seats, and pedestrian plazas.

Although they might seem like one of the least ecological features of cities, parking lots hold high greening potential. Communities can require that tree canopies (or photovoltaics) shade the paved surface and that stormwater be infiltrated on-site through pervious pavement or planted swales at the pavement's edge or between rows of parked cars.

In the long run, however, parking areas should be viewed as land banking—opportunity sites for future parks, gardens, playgrounds, or buildings. In any vibrant urban area, surface parking lots disappear when the land becomes too valuable for this low-intensity use. Parking is then integrated into structures and surface lots are redeveloped. If autonomous vehicles become widespread, less parking may be needed in central cities, since fewer people may drive their own vehicles and need to store them nearby while at work or home. Improved land use balance, pedestrian-friendliness, and transit service within communities may also help reduce the need for motor vehicle use and parking.

Encourage Green Roofs and Walls

Green roofs have received lots of attention in recent years as a way to add vegetation to cities, improve energy efficiency of buildings, slow stormwater runoff, and provide attractive outdoor spaces for building users. There are two types: "intensive" green roofs that people actively use, and "extensive" green roofs that are generally not publicly accessible but are put in place for other benefits such as energy conservation.

In temperate climates green roofs typically use a variety of grasses and forbs (flowering herbaceous plants) and need little if any irrigation. In drier terrains designers often employ succulents such as sedums, which need little irrigation or irrigation only to establish themselves. Potentially awnings or solar panels can shade rooftop plantings in hot and dry locations. Transparent and opaque solar panels have been invented that might be useful in such circumstances.[24]

Patios with plants, raised-bed gardens, and even small trees in containers are another green use of roofs. Urban dwellers have traditionally created a variety of such rooftop amenities throughout the world. One main problem is that building owners or maintenance staff often do not allow residents access to the roof. This is a problem that cities could potentially address through code requirements for rooftop access if basic safety steps are taken.

Green walls are a bit more challenging. The vegetation must be provided a support structure affixed to the wall, provided with a growing medium as well as irrigation in dry climates, and maintained by workers despite the vertical heights. The sun will bake south-facing and west-facing walls in many climates, so designers may need to use extensive irrigation or drought-tolerant succulents. North-facing walls will often be in shade in the Northern Hemisphere, requiring a very different plant palette.

Still, green walls are an exciting opportunity to add a highly visible natural element to urban locations. They often impart a certain shock factor, since viewers don't expect to see plants covering buildings. Ivies have traditionally grown on the walls of brick buildings in certain temperate-climate locations, but these have low habitat value and are less interesting than green walls with a more varied plant palette. If it is not possible to put plants right on the wall, designers can often train ground-based vines up trellis structures in front of walls to provide a vertical green feature. Vegetation-covered trellises and walkways at ground level can be part of urban greening as well.

Reduce Hardscape

The amount of asphalt, concrete, and roofing in a city is inversely correlated with a number of types of ecological performance. The more hardscape, the more communities overheat in the summer from urban heat island effects. As mentioned in chapter 2, low-income neighborhoods and those housing people of color (often previously redlined communities) typically have less tree coverage and experience greater heat, so there is an environmental justice component to heat exposure.[25] Dark-colored asphalt not only increases temperatures during the day but soaks up solar radiation and re-radiates it at night to keep cities much warmer than surrounding territories. This translates into lower human comfort and increased use of air conditioning, thus leading to higher electricity consumption and GHG emissions. Global warming will only aggravate this phenomenon.

Hardscape also leads to rapid, high-volume runoff of stormwater. Instead of sinking slowly into the ground as previously, rainfall immediately runs off building roofs and pavements into municipal storm drain systems. From there ecological damage can occur in a number of ways. Many cities have combined sewage and stormwater systems, meaning that sudden influxes of stormwater go to the municipal sewage treatment plant and often overflow holding tanks and ponds, contaminating the nearest body of water with human waste. This is one reason urban rivers are unswimmable and city beaches are often closed because of elevated E. coli levels after storms. Alternatively, if cities have a separate stormwater system, heavy rainfall often flushes stormwater into the nearest stream or river. Those riparian systems never had to deal with such rapid pulses of stormwater under preurban conditions, since rainfall stayed on-site far longer and much of it percolated into the ground locally. The sudden, rapid stormwater surge erodes stream channels and washes away vegetation and habitat. In many urban areas one can see that channels are now far deeper than before urbanization and are bereft of small plants.

Reducing hardscape at every opportunity is the most basic way to address these problems. Designers and engineers can often reduce the dimensions of streets and parking lots to use the minimum amount of asphalt or concrete necessary. They can also often specify permeable versions of asphalt and concrete. In these, the smallest particle sizes, known as "fines," are dropped out of the aggregate mix when producing the material, leaving larger pores that water can flow through.[26] These alternative paving materials are not perfect; permeable asphalt in particular is often not as strong as the traditional kind. Also, particulates carried by stormwater can clog the air spaces within permeable materials. Quarterly maintenance by means of sweeping, vacuuming, and high-pressure water-jet cleaning is likely to be needed to keep permeable pavements fully functioning.[27] Such maintenance will increase costs; on the bright side it can also create green jobs.

Promote Climate-Appropriate Landscapes

Private yards are an enormous frontier for urban greening. Historically residential yard design—consisting mainly of lawns in temperate climates—has often been of low ecological value and requires large amounts of water, fertilizer, and herbicides. Including even a small number of well-chosen native species can provide habitat for bees, butterflies, birds, and insects. In the US, the National Wildlife Federation has run a backyard habitat program since 1973 (now rebranded "wildlife gardens") encouraging homeowners to increase the habitat value of their yards.[28]

To save water, many cities in the southwestern US either prohibit turfgrass or pay homeowners to remove existing lawns and replace them with more drought-tolerant plants. In the early 2000s the City of Albuquerque paid one of us $0.80 per square foot to rip out the grass lawn of his house in New Mexico and xeriscape—an $800 incentive. The resulting yard was more water-efficient, of higher habitat value, and more varied and interesting to look at.

Those who live in apartment buildings may have little control over the exterior space around them. But there is often something that can be done, if only to add planters to a porch, balcony, or entrance. In older neighborhoods of Japanese cities, where homes fill almost the entire lot, residents still manage to place plants in pots along the street frontage. This small step helps personalize and green these neighborhoods. Cities can also develop programs encouraging landlords to green spaces around their buildings.

Expand Urban Agriculture

Community gardens (in North America) and allotment gardens (in Britain and northern Europe) date back more than one hundred years and provide urban dwellers ways to both grow produce and connect to the Earth. During the Second World War residents of many countries also grew produce in their yards as "Victory Gardens."[29] Community gardens provide North American urban dwellers with relatively small plots for gardening, often one hundred square feet or less. Allotment gardens provide European urban residents much larger garden spaces that often feature a small shed or cottage. Perhaps the ultimate example is the Russian dacha, which during the Soviet period was a main perk given to factory workers. Workers and managers would often receive identical plots of up to four thousand square feet (372 square meters) in a dacha development outside Moscow or other cities. Families would construct cottages, ride local trains out to the dachas on weekends, and live there much of the summer, growing produce that they could eat year-round.

In recent decades many cities have made new efforts to reconnect with local agricultural production. These initiatives include urban farmers' markets, schoolyard gardens, community-supported agriculture (in which groups of urban residents contract with local farmers for produce), backyard gleaning programs, and "farm to fork" educational campaigns. Young urban farmers and community residents are farming vacant lots and rooftops, while other green entrepreneurs have sought to more effectively distribute produce from private yards. Social and racial justice advocates are connecting urban agriculture with the need for food justice, Black land ownership and control, neighborhood self-determination, reclamation of land, and environmental justice.[30]

Urban agriculture isn't simple. It is easy to romanticize the idea of locally grown food without realizing the difficulties of market production in urban areas, including expensive land and a lack of small-scale distribution networks for local produce. Urban farmers must compete with cheap conventional produce grown with the aid of subsidized, fossil fuel-based fertilizers. Hopefully such subsidies will be eliminated, evening the playing field. Urban farming runs into other problems as well. There is sometimes a question about who owns the food grown on these plots. Planting fruit trees along public streets may lead to messy sidewalks, which the city's public works department won't be happy about. There are also questions about urban soil quality and safety. Few urban dwellers have either time or expertise to farm small gardens efficiently or to use the bounty of even one or two high-yielding fruit trees whose produce must be picked and processed in the short period of time it is ripe. Making urban agriculture more productive is likely to require substantial training, professional assistance, government or nonprofit support, and/or lifestyle change. If the goal is to promote food justice, it is important to note that disadvantaged communities should also have good, accessible, and affordable grocery options available, rather than having to grow food themselves.

Many city residents have also begun keeping chickens, bees, rabbits, or other domestic animals. Chickens are far-and-away the most popular species, kept for their eggs. But beekeeping (often on urban roofs) is increasingly popular as well. Local governments have had to review their codes to mesh desires for such animal husbandry with public health and safety. Chickens are often allowed; roosters aren't. The prevalence of more traditional companion animals such as dogs has increased as well, necessitating new dog parks and rules about cleaning up after one's animal.

Reduce Light and Noise Pollution

Although it is something that most people rarely consider, light pollution is yet another way in which urban areas become less natural. Unshielded streetlights and billboards make the urban night sky a murky haze and can eliminate one's ability to see stars over cities. There are actually three main kinds of light pollution: general "sky glow" over urban areas, "light trespass" in which unwanted bright light from one property shines onto an adjoining property, and "glare," in which bright light causes visual discomfort to individuals. Excessive light pollution may disrupt circadian rhythms in both humans and animals, possibly leading to sleep problems and depression. It also represents a waste of energy.

Since the late twentieth century a number of jurisdictions have sought to counter these impacts, usually by requiring that outside lights be shielded so that light goes only toward the ground or that such lighting minimize energy use. Examples include Arizona, which first required shielded outside lighting in 1986, and New Mexico, which prohibits outdoor lighting of recreational facilities after 11:00 p.m.[31] Noise pollution is a related detriment to urban landscapes that feel green and natural. Better regulation of motor vehicle noise—and the shift toward electric vehicles—will help greatly with this.

Avoid Green Gentrification

To avoid the green gentrification issues mentioned in chapter 4, urban greening strategies need to meet the needs of all population groups, not just well-off white residents. They also need to be combined with strategies to undo environmental injustice and racism, reduce social inequality, provide affordable housing, protect renters from displacement and low-income homeowners from increased property taxes, and honor historic neighborhood cultures and character. Cities need to actively involve lower-income residents in designing urban greening programs as part of an overall package of strategies to both improve a neighborhood and protect existing residents.

CONCLUSION

In an era of enormous urbanization worldwide, it is essential to green cities and make them biophilic so as to provide residents with everyday connection to nature. Greening cities doesn't just mean adding vegetation; it also requires reducing much of the noise, traffic, pollution, danger, and environmental injustice of existing urban environments. Greening strategies can meet multiple urban sustainability goals at once. For example, using green street design to slow traffic improves street safety and aesthetics, makes streets more desirable for bicyclists and pedestrians, helps cool the city, improves habitat value, and allows for on-site drainage of stormwater. People living in these communities may have the protective amenities they need to weather future challenges like the Covid-19 pandemic and rising urban temperatures. Such urban greening initiatives can help make communities far more attractive, resilient, and equitable places to live in the twenty-first century.

In the end, greening cities is not just about reintegrating nature into human communities but about promoting a different set of values within a new, sustainability-oriented social ecology. Urban greening is the physical manifestation of our desire to value the Earth and not just our own pragmatic or selfish needs. It is one way to express our love for the ecosystems in which we live. It is also a way to nurture our own spirits and add joy and delight to everyday life. These new motivations represent a twenty-first-century mindset very different from that held by previous generations who created BAU urbanism.

9

How Do We Reduce Our Ecological Footprints?

How much should each of us do in our personal life to bring about sustainable societies? Yes, we know we should turn off lights when we leave a room, recycle, and not waste food. We should drive and fly less. We should buy less, live in smaller homes, put on a sweater and turn the heat down. We should probably organize family and friends to do like-wise. But sometimes we get tired of hearing these warnings, and we wonder, Do such small actions really make a difference?

The same question occurred to John Javna, author of the 1989 #1 US best seller *50 Simple Things You Can Do To Save the Earth,* about five years after he had written that book. By that time *50 Simple Things* had sold five million copies and been translated into twenty-three languages. Its daily eco-actions included composting kitchen scraps, using cloth diapers, snipping six-pack rings so they wouldn't choke sea creatures, and choosing paper over plastic bags at the grocery store. The implica-tion was that cumulative actions on a personal scale would be enough to "save the Earth."

Javna's original list of "simple things" proved useful to some people but was ridiculed by others. Some recommendations turned out not as simple or straightforward as they appeared. For example, reusing cloth diapers may save landfill space and materials, but washing them uses lots of hot water and soap, not to mention requiring more labor from parents. More importantly, Javna came to realize that the notion that simple actions were enough to save the planet was deeply flawed.

Sustainability isn't really so simple. Enormous political and economic forces work against it, and unless we collectively address those structural problems, individual actions won't be nearly enough. He took the book out of print in the mid-1990s, grew cynical, and for a while stopped doing simple things himself.[1]

In 2006 Javna's thirteen-year-old daughter asked him why he wasn't doing more of the daily actions he'd written about. Chagrined, he rethought his approach and in 2008 released a completely rewritten version, emphasizing that big-picture, structural changes are crucial to sustainability as well as small-scale steps and that the two are deeply linked. This book presented a new list of "50 Simple Things," developed in conjunction with environmental organizations such as Friends of the Earth, Rainforest Alliance, Sierra Club, the Alliance to Save Energy, and the Institute for Local Self-Reliance. Each step tried to make connections between individual actions and social change. Javna emphasized the importance of everyone getting involved in any way they can, saying that "it doesn't matter which issues we choose to work on—big or small—because they're all connected."[2]

Big-picture changes such as those discussed elsewhere in this book are necessary to counter BAU. Those structural transformations need to be emphasized. But Javna's original book had merit as well. To create sustainable cities or get to a sustainable society, each of us will need to move toward ways of living that make that possible. The UN's Sustainable Development Goal (SDG) 12, "Sustainable Consumption and Production," seeks to get at this dimension of sustainability through an emphasis on meeting basic needs and managing natural resources, although without quite saying that overconsumption should be reduced.

Moreover, not thinking in terms of personal behavior skews policy debates in misleading directions. For example, as we saw in chapter 1, leaving household consumption and diet out of analyses of GHG emissions, as is usually done in economic sector-based GHG accounting, leads to misleading statistics for many societies. It also absolves us of responsibility and agency. Emissions and pollution from products people in wealthy countries buy and food they eat are in effect exported to other parts of the world

Behavior change questions have no simple answers. Addressing them means questioning personal and societal values and potentially changing the ethics and priorities of daily life. It also means questioning the assumption that capitalist economies must be based on ever-expanding growth in personal consumption. Finally, it means connecting the personal and the political. How do we "walk the walk" on topics such as climate change—by leading carbon-neutral lives—as well as "talk the talk"? And how do we help governments create policies that assist us to "walk the walk"? After all, doing the right thing should be also be the easiest choice under a more sustainable political economy.

The subject of behavior change raises important questions of equity. It would be manifestly unfair for rich nations to say to poor ones, "Maybe you should just stay at your current low level of consumption because the world has a climate and ecological crisis." Ways for people in developing nations to enjoy higher standards of living while living low-impact life-styles must be found, and wealthy societies must take responsibility for the negative impacts of their own excessive consumption and reduce those accordingly. This is what climate justice entails. Some would argue that major financial and technical assistance from the developed to the developing world is necessary to assist in this transition, or even that the former should make reparations to the latter for the climate damage it has caused. Within a given society, acting sustainably should not be a luxury that only the wealthy can afford. Eating foods without pesticides should not be a luxury or a badge of honor. Driving in cars that don't emit emissions or living in eco-friendly homes and communities should not be a luxury. Economic and political logics will need to change because sustainable behavior should not be a luxury; it is a necessity.

CAUSES OF EXPANDING FOOTPRINTS

The concept of everyone having an "ecological footprint" representing the impact of their lifestyle and behavior on the planet is relatively recent. It arose in the 1990s through a collaboration between William Rees, a professor at the University of British Columbia, and graduate student Mathis Wackernagel.[3] However, the concept has historical roots in ecological notions of "carrying capacity," the number of organisms of a particular type that a given habitat can sustain. It also draws upon research on resource depletion, life-cycle impact of products, and global limits to growth.

The ecological footprint metric attempts to quantify the biologically productive area of land it would take to support a given person or society. In reality this estimate is only approximate and incomplete. Many types of human consumption cannot be translated easily into land area. Although efforts have been launched worldwide to try to measure ecological footprints, such as by the nonprofit Global Footprint Network

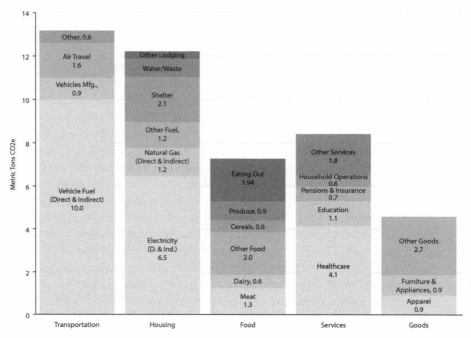

FIGURE 11. The average US household's carbon footprint in 2015. It will be essential to reduce household carbon footprints to almost zero within a couple of decades. Footprint components vary from place to place. A household footprint calculator is available at https://coolclimate. berkeley.edu/calculator. (Christopher M. Jones/Cool Climate Network, personal communication, August 2020.)

(figure 11), the device is most useful as an educational tool to get us to think about impacts of our lifestyles. An online ecological footprint calculator that individuals can use is available at www.footprintcalculator .org. Upon completing such an exercise, the average resident of developed nations finds that two, three, or even four or more Earths would be required to support their lifestyle if shared by everyone globally. This should be a wake-up call for the world.

A few profound changes during the past century have contributed to the current outsized impacts of our lifestyles on the planet. The first of these is diet. Consuming large amounts of meat and dairy products, often viewed as a sign of wealth in developing countries, has enormous impacts. One study of American GHG emissions estimates that meat, dairy, and processed foods account for almost six metric tons of GHG per capita.[4] Globally, the livestock sector uses one-third of the world's cropland just to produce feed for animals and fowl, as well as one-third of available fresh water.[5] Livestock are responsible for about 12 percent of the world's GHG emissions through methane emissions from manure and ruminants as well as the embodied emissions in feed.[6] Using so much land for animals likely contributes to food scarcity and hunger for hundreds of millions of people, since these individuals do not have access to land to grow the fruits, vegetables, and grains they need for subsistence.

Processed foods—items that have gone through extensive commercial production and are typically highly packaged—also have high environmental impact as well as frequent negative health implications. These products often use corn syrup, refined sugars, artificial food coloring, and other ingredients that are both unhealthy and environmentally destructive to produce. Packaging alone includes substantial amounts of plastic or paper, each of which generates impacts all along its supply chain.

Growing mobility is another factor that contributes to industrial lifestyles' impact on the planet. A century ago a vanishingly small percentage of the world's population drove to work every day, for the simple reason that few cars existed as well as few good roads to drive them on. Most people worked in or near the home or else walked to work. The picture is radically different today. Vehicles cover the planet. By 2040 the number of cars in the world is likely to reach two billion (almost doubling).[7] Not only have motor vehicles polluted the air and become the leading GHG source in many places, but their influence has completely changed the form of cities and towns and the character of rural environments as well. They have allowed radical decentralization of communities, with people living in larger houses on larger pieces of land and consuming correspondingly more.

Connected with both of the above transformations is the rise of the fossil fuel economy, beginning with industrial use of coal in the nineteenth century but expanding greatly in the twentieth with the rise of motor vehicles, industrial agriculture, and petrochemicals. Before the second half of the twentieth century everyday products such as chairs, tables, kitchen utensils, and clothing were made from wood, metals, clay, or natural fibers, all of which can be reused, recycled, repurposed, or decomposed back into soil. Beginning in the 1950s, plastics and other products of the petrochemical industry made possible an explosion of inexpensive household objects based on new materials, but at high environmental cost. In the process the engine of capitalist economics got seriously off track, promoting consumption for its own sake and using strategies such as planned obsolescence to keep consumers spending money.

An argument can be made that cheap, plentiful fossil fuels made industrialization possible and led to benefits such as modern medicine, hygiene, telecommunications, and labor-saving appliances that were particularly important for women, assisting their entry into the labor force. Water power and use of wood or other biological materials for combustion probably wouldn't have been sufficient. But for many years alternatives to fossil fuels have existed. In particular, electricity generated by sun, wind, water, and geothermal energy could power a modern economy. So if fossil fuels were in fact necessary as a bridge to modernity, that need is now gone. The dream of a Green New Deal is possible if we have the political will to make it happen.

Other factors as well have increased ecological footprints over the past century. The increasing size of houses in many countries, with their attendant needs for heating, cooling, and furnishing, contributes to higher impacts. So too does inequality, since the rich overconsume and the poor both seek to emulate them and cannot afford to do so in an environmentally responsible manner, for example buying less efficient and more polluting vehicles. Suburban sprawl has required more motor vehicles and more driving. The advent of specific technologies such as air conditioning and electronics has made a profound difference in consumption. Air conditioning, for example, has allowed great expansion of communities in very hot climates (places that should probably not have been developed), at an enormous cost in terms of energy use. Gaining awareness of such trends, both in society at large and in our personal lives, is a starting point for change.

Human needs are in large part socially determined. A couple of generations ago, many Americans lived relatively well with a small fraction of today's material consumption. People in other cultures are now doing the same. Some of those countries have higher quality of life, longer life expectancy, and more positive happiness ratings on international surveys than high-consumption societies like that of the US. So human needs are quite flexible. During the Covid-19 pandemic many Americans learned that they could live very well without many things they typically consumed.

Continuing and expanding such lifestyle changes is part of the challenge. At the same time, many goods could be produced locally or regionally rather than transported from distant parts of the world. Many could be made from wood, metal, or other more traditional materials rather than plastic. Many could be sold with less packaging. Such changes could further reduce the ecological footprint of average lifestyles.

ALTERNATIVE ETHICS

How can individuals live more lightly on the planet? A few basic philosophical shifts can help (table 9). These ethics can be articulated by leaders at all levels, reinforced through educational systems, promoted by social marketing campaigns, and emphasized by parents.

Live Simply

Even within capitalist society—and for millennia within certain religious traditions—many people have followed an ethic of voluntary simplicity. Sometimes this results in a minor paring back of possessions. Sometimes it leads to a wholesale rejection of consumption, with individuals buying nothing new, eating very simply, and downsizing living accommodations to a bare minimum. Recent interest in tiny homes is one manifestation of this ethic. These structures usually of less than four hundred square feet (thirty-seven square meters) are the opposite of trophy homes and McMansions and perhaps represent a psychological reaction against widespread overconsumption. They can allow people to afford to live in their preferred communities and save money.

Voluntary simplicity can extend to all forms of consumption, for example diet, air travel, and personal services, and is often combined with a corresponding respect for nonmaterial things in life. Rather than gaining meaning through things, people seek to appreciate daily forms of experience, including experiencing of other people and the natural world. It is a mindset that emphasizes awareness of the impacts of our consumption and respect for other people and the planet.

Eat Plant-Based Diets

"Eat Food. Not Too Much. Mainly Plants." This quote from Michael Pollan, the journalist who has written widely on food systems, summarizes what is both a new ethic for dietary consumption and a reasonable summary of how most humans ate until the second half of the twentieth century. Meat was a luxury in most places until the modern era. Packaged food was rare or unknown. Substances like high-fructose corn syrup, routinely added to many processed foods today, had not yet been invented.

Many of these new, modern foods are bad for the planet, producing GHGs, leading to land degradation and deforestation, and requiring

TABLE 9   STRATEGIES TO REDUCE ECOLOGICAL FOOTPRINTS

Strategy	Description
Live simply.	Adopt an ethic of voluntary simplicity with less consumption, reuse of clothing and possessions, smaller homes, less travel, and celebration of nonmaterial things in life.
Eat plant-based diets.	Avoid meat, dairy products, and packaged or processed food.
Make kin, not population.	Develop families of choice rather than large biological families.
Give rather than take.	Emphasize values of caring, community, and service to others rather than acquisition and consumption.
Develop regulation and incentives to reduce consumption.	Make it easy for people to "do the right thing" by limiting unsustainable choices and rewarding them for making good choices—for example, by requiring that appliances be energy-efficient and offering rebates for purchasing the most efficient ones.
Internalize externalities.	Make sure the real social and ecological costs of lifestyle decisions are captured in their price; incentivize more appropriate choices.
Create social marketing campaigns.	Create educational and marketing campaigns aiming to change individual thinking and social discourse around consumption and the reduction of ecological footprints.
Develop communities of support.	Help individuals create peer support networks; organize community events and publicity around sustainable behaviors and values.
Take advantage of crises.	Leverage crises and turning points for maximum educational gain, value shift, and lifestyle change.

fossil fuel–intensive mechanized agriculture. They are also bad for people's health. Use of high-fructose corn syrup, for example, is correlated with increases in type 2 diabetes.[8] High consumption of red meat, especially processed meat, is associated with increased risk of stroke, heart disease, and breast, colorectal, and prostate cancers.[9] Traditional diets were both healthier and more sustainable. If everyone ate a Mediterranean diet, for example, with lots of nuts, beans, grains, vegetables, fish or chicken once a week, and red meat even less often, the impact would be equivalent to taking a billion cars off the road.[10] Of course, such foods—organic, locally produced, minimally processed—must be affordable and easily available to all urban residents.

Changes are needed both in our individual consumption practices and in global food systems. Structural changes in societies are necessary to reform food systems—for example, the elimination of subsidies for large corporate farms producing commodity crops like corn and soybeans (both used to feed cattle and hogs, thus subsidizing meat-heavy diets). However, at the city level many smaller actions can help change the system. Food stores can emphasize bulk sales of grains, nuts, pasta, beans, vegetables, and fruits by putting these at the front of the store and can eliminate many processed foods altogether. In New Zealand, grocers came up with the idea of a "food in the nude" campaign through which they pledged to market fruits and vegetables without plastic packaging.[11] Meanwhile, neighbors can share surplus garden produce. Restaurants and grocery stores can donate surplus food to homeless shelters. Gleaners can scavenge excess fruit from trees of participating households and make this available to schools or soup kitchens. Food sellers can offer classes helping customers learn new ways to cook healthy food. The Davis (California) Food Co-op, for example, runs a teaching kitchen offering dozens of classes about healthy cooking and eating.

Make Kin, Not Population

Population is a much-overlooked dimension of sustainability. Politicians and activists have generally avoided the subject because of unsavory associations with eugenics, forced sterilization, racism, and anti-immigrant sentiment. But the size of populations matters enormously as societies move into greater affluence. It is impossible that eight, nine, or ten billion people can share anything approaching current industrial-country lifestyles without overwhelming the planet. An ethical mandate to reduce average family size seems needed.

Luckily, the natural tendency is for birth rates to fall as women receive greater educational, economic, and reproductive freedom (something that is often under threat). This is emphasized by SDG 5, Gender Equality, even though the SDGs generally avoid the controversial topic of population control. Domestic populations within countries such as Japan, Germany, and Italy are already declining (though immigration offsets such declines in some places). But they must fall faster if humanity is to achieve carbon neutrality by midcentury. Population decline has its own challenges—for example, how to support a growing elderly population with a smaller workforce and how to retain the vitality of cities and towns with fewer people. But these can be addressed. In many cases, immigration can help balance population decline—if existing residents are willing to accept people with characteristics different from their own. Societies can also develop "managed retreat" programs in which small towns or neighborhoods in ecologically sensitive areas are bought out for purposes of environmental restoration, with remaining populations relocated elsewhere.

Many people historically have wanted large families to provide security and comfort in their advancing years. However, good social safety net programs can address the security concern, and as Adele E. Clarke and Donna Haraway have argued, the human need for community can be met in the future through a strategy of "making kin, not population," that is, by creating families of choice, not of blood connection.[12] Few would want to set limits on family size, as China did in the late twentieth century, but greater awareness of the impact each child has on the world is important, as are reproductive rights for women and attention to how societies can care for people the world has now.

Give Rather Than Take

Perhaps the most basic personal shift will be in the balance between individual rights and collective responsibilities. Laissez-faire capitalist societies emphasize individualism to an extreme. Self-centeredness and narcissism frequently result, illustrated most famously by Donald Trump. These qualities keep advantaged groups within societies from examining the structural elements that maintain their own privilege and keep others from the same rights, opportunities, and security. The Black Lives Matter movement has helped point out these structures with regard to race and white supremacy.

To move toward more sustainable societies, the goal must be a society of care and community and healing. Personal values must emphasize giving back to the world, not profiting from it. Citizens must acknowledge their responsibilities to one another and the planet. Institutions will need to change to support communitarian values and service, for example by paying teachers as much as financial advisers, and perhaps by requiring some amount of public service from everyone.

A sustainability ethic in daily life would combine Golden Rule formulations (treating others as we would want to be treated ourselves) and Aldo Leopold's Land Ethic (treating the planet in ways that improve the biotic whole). Most simply, it might tell us, "Make every action improve the well-being of both people and the planet." Along with this must go greater individual awareness of the complex ways in which our actions affect the systems around us, and of alternative ways to make those impacts more positive.

This giving-rather-than-taking ethic applies particularly in the work world. So many professions are oriented toward simply making money or actively exploiting others for personal gain. Capitalist economics presents a range of rationalizations for such behavior, such as the invisible hand of the market (the belief that "free" markets by themselves can ensure collective welfare, though because of market imperfections they are highly unlikely to do so). A related myth is that wealth will trickle down to the poor. A third is that concentration of wealth is necessary for motivation (humans can be motivated in many other ways, such as by concern for others and finding meaning in life).

Under a radical sustainability approach, societies might proactively abolish professions that prioritize taking rather than giving. Many forms of financial speculation come to mind. But people need to actively make choices to do work that gives to the world rather than taking from it, in some cases looking beyond the range of alternatives that their home community presents. Educational systems, the media, and the political economy need to encourage those decisions. A life well lived needs to be one where people actively contribute to the world.

HOW TO CHANGE BEHAVIOR?

Various behavior change theories suggest ways to encourage sustainability-oriented lifestyles and behavior. None is sufficient in itself, and our understanding of how societies can make major value shifts is still in its infancy. But each perspective has insights to offer.

One main school of thought is social cognitive theory, associated with Stanford psychologist Albert Bandura. This approach emphasizes external factors such as social models, stimuli, and reinforcements that interact with an individual's self-efficacy (inner ability to achieve goals) and self-reflection to produce behavior change.[13] Such a perspective points to the importance of the media, entertainment, peer groups, and role models in bringing about lifestyle change. Following this approach, bringing about more sustainable individual lifestyles might require structural changes in media (for example, banning advertising of certain categories of products as well as misinformation), leadership from political and cultural figures, and peer group training in sustainable living during K-12 education and even in higher education. Probably many other strategies as well would help give individuals the cognitive skills, self-awareness, and self-confidence needed for lifestyle changes.

Another approach is the so-called theory of planned behavior, which emphasizes the individual's intentions, attitudes, beliefs, and subjective norms.[14] Although overlapping to some extent with Bandura's approach, here the focus is on inner motivation and people's reasoned evaluation of alternatives. A main implication is that people actively choose behaviors rather than passively falling into them. The belief that one can make a difference is thus important in changing behavior. It becomes crucial to present information in ways that can help individuals understand the difference they can make and feel good about it. For example, a low-carbon household program might develop a website giving a family a clear chart of their GHG emissions by month and year based on input of simple information. They could then see on a daily basis how they had lowered their carbon footprint. Their local government might also award prizes for the greatest household GHG reductions and post these households' names on municipal websites.

Other researchers present different approaches to behavior change. James Prochaska and Wayne Velicer propose a six-stage model of change: precontemplation, contemplation, preparation, action, maintenance, and termination.[15] Strategies for behavior change might target one or more of these stages. For example, informational campaigns might aim to get low-consumption living on people's radar to begin with (the precontemplation stage) through advertising and exhibits in schools and public spaces. Authorities might then provide resources for people actively seeking to reduce the carbon footprint of their diet or making their homes zero-net-energy (the contemplation stage), and might organize support groups for long-term maintenance of such behaviors (the maintenance stage).

Stanford behavioral scientist B. J. Fogg emphasizes the role of motivation, specifically pleasure and pain (immediate motivators), hope and fear (more long-term motivators), and social acceptance and rejection (active across multiple time frames).[16] He argues that for new behaviors to occur, the individual must be motivated, able to do the behavior relatively easily, and triggered by some event or reminder.[17] For example, the real-time fuel efficiency displays on the dashboards of many hybrid motor vehicles provide immediate motivation (I'm saving gas and GHGs!), an easy action (going a bit lighter on the accelerator), and constant triggers (a bar going up or down on the display to show gas consumption in any given moment, or even on certain models of Prius a constant grade on the scale of 0–100). Social acceptance is also involved, at least among certain peer groups, since hybrid or electric vehicle owners are often respected for their environmental commitment. Similar types of dashboard displays might be created within homes to encourage residents to turn down the air conditioning in summer or to check to be sure that all unneeded lights and appliances are off.

Finally, social marketing strategies use ideas from advertising to bring about positive social change. These approaches have been applied most prominently in public health campaigns to get people to smoke less, eat healthier diets, and live actively and exercise. Social marketing campaigns have also been undertaken around family planning, HIV prevention, heart disease prevention, skin cancer prevention, disaster preparedness, and hygiene and sanitation. They could be used to assist with virtually any sustainability-oriented behavior change.

Canadian psychologist Doug McKenzie-Mohr has developed a process called community-based social marketing (CBSM), through which researchers use focus groups and surveys to discover barriers to change and then develop specific strategies to surmount these barriers. Such strategies can include particular types of messaging, incentives, ways of making actions more convenient, and steps to change social norms.[18]

Many authorities warn against using guilt and shame to change behavior. Yes, these can motivate people towards a desired direction in the short term. However, they often engender resentment and are counterproductive in the long run. So it becomes important to make sustainable lifestyles fundamentally more satisfying, meaningful, and fulfilling—and for society to reward them accordingly. We also need to make sustainability ubiquitous and easy to choose through policies that support sustainable choices. It becomes essential to change institutions within social ecologies so that new ways of living become more possible, affordable, and accessible, not just for a few to pursue, but for everyone within society to choose.

TOOLS TO GET THERE

Develop Regulation and Incentives to Reduce Consumption

As with many other sustainability challenges, public sector codes, laws, regulations, and incentives are necessary starting points. Energy efficiency codes decrease the energy used in homes without occupants having to lift a finger. Phasing out gasoline-powered cars and trucks will spur electric vehicle innovation. Tax credits and rebates encourage people to more actively choose energy-efficient appliances. Tax deductions for charitable giving encourage individuals to give money to community-serving organizations.

Although it sometimes has a bad name, top-down public sector regulation is essential in order to change individual ecological footprints. Few of us have time to research the environmental and social impacts of every purchase. Products' impacts may also be hidden through their life cycle. Moreover, the market alone may not result in *any* good choices for us. So the public sector must set sustainability standards, ensure that harmful products are banned, and in some cases restructure markets to promote sustainability transitions. For example, automakers brought electric cars on the market in the 2010s in large part because regulation in places like California required them to. Publicly mandated product labeling is also essential for consumers to be able to make sustainable choices.

Internalize Externalities

A second main set of strategies uses economics to ensure that the real social and ecological costs of our lifestyle decisions (their negative externalities) are captured in the price we pay for them. Air travel, for example, is probably vastly underpriced compared to what the price would be if the environmental cost of the GHGs it generates were included at a price of, say, $100/ton. Incorporating externalities into the price of goods and services has long been a goal of environmental economists. Although techniques for estimating the cost of pollution and other externalities are at times crude, mechanisms do exist to estimate such cost and include it within the price. Taxes and fees are the most obvious way to do this. A carbon tax seeks to incorporate the costs of climate change into the price of fossil fuels; a fishing license may reflect costs associated with the depletion of fish species and may generate revenue that can be used to manage a fishery.

In a different way, economic mechanisms such as high parking charges can be used, not as a true reflection of the cost of parking, but as an incentive for behavior change. It may be in the interest of communities and the environment to have people leave their cars at home and bike, walk, or use public transportation to travel downtown. Calculating the price of these social and environmental benefits in economic terms would very difficult, but setting high prices on vehicle use can help promote behavior change, particularly if other more sustainable transportation options are supported.

Some economic strategies for behavior change are effective, others less so. Parking charges are among the best ways to get people to drive less. Gas taxes are less effective. North American drivers complain a lot about any increase in gas prices—and local TV news programs are fond of running stories about this—but motor vehicle owners go ahead and fill their tanks regardless (often because they still need their cars for commuting), shifting their driving patterns relatively little. Instituting a gas tax and incentivizing public transit use and transitions to electric vehicles may still be important to do. But as with all policy change, it is good to be aware of the equity impacts and potential trade-offs ahead of time.

The philosophy known as limitarianism would have the public sector use the tax system to eliminate excessive wealth and consumption. Advocates promote the view that no one should hold surplus money above what is needed for a healthy and flourishing life and that any excess should be taxed away.[19] This would keep ecological footprints limited and would also generate massive amounts of tax revenue for constructive social purposes. In a world where one man (Jeff Bezos) is worth more than $200 billion, there is much to be said for this point of view.

Create Social Marketing Campaigns

More individually oriented behavior change strategies exist as well. Prominent among these are social marketing campaigns. These messaging programs are carefully designed to educate people and change individual thinking and social discourse around particular topics. The antismoking, anti-drunk-driving, and healthy living/antiobesity campaigns of recent decades are prime examples (the first of these uses taxes extensively as well).

Social marketing campaigns can be effective, especially if done on a large scale and integrated across different types of media and institutions (for example, schools, public health facilities, workplaces, and cafeterias). Ideally, individuals receive the same message repeatedly from many sources, and the message is clear, simple, memorable, and doable. Campaigns can use peer influence to reinforce behavior change. For example, active living campaigns have at times distributed flyers and fridge magnets suggesting that individuals "Take a Friend for a Walk." This is a way of encouraging people to exercise, making the exercise more fun, and also encouraging the friend to adopt the same behavior. (Of course, this assumes that cities also take the necessary steps to ensure that streets, trails, and public spaces are safe and pleasant, so policy change matters as well.)

At the same time, public policy can seek to limit counterproductive messaging that individuals are exposed to. In this way many nations have limited advertising for cigarettes or alcohol. Many have also considered limits on the marketing of soft drinks and packaged foods to children. (Economic disincentives such as taxes on cigarettes and soda can go hand in hand with advertising bans and social marketing campaigns.) Limiting television and online advertising may be desirable to encourage less materialistic lifestyles. Tackling the impacts of social media on consumption patterns may also be necessary. Alternatively, a ban could be limited to material products with high carbon footprints.

There are downsides as well as potential benefits to social marketing campaigns. They may come across as heavy-handed—"Big Brother" wanting us to do things differently. They may lead to resentment or resistance, as when young people consciously drink or smoke in defiance of society's prohibitions or people refuse to wear masks during the Covid-19 pandemic. They may involve substantial costs to government and may put politicians at risk of being seen as representatives of the nanny state. However, commercial interests have long known the power of advertising. Why should they be allowed this power while the public interest isn't? Carefully designed public communications and social marketing campaigns are likely to form a central part of any move toward more sustainable cities and societies. All of us will need to develop better skills in environmental communication, for example modeling sustainable living and advocating for policy change to make it more accessible and attainable rather than simply preaching about it. At the same time, we need to recognize that truly sustainable and equitable societies make sure that everyone's basic needs are met, connecting concerns for people's well-being with that of the earth.

Develop Communities of Support

Behavior change will be difficult if each of us is alone in our quest to live more responsibly or if individuals are ridiculed for being tree huggers or if sustainability it seen as "elitist." Peer support and networks of others doing the same thing will help greatly. The Internet is great for this, since it allows new networks around a particular topic to be set up relatively easily and enables people to find others interested in the same thing through searches. Freecycle networks, ride-share sites, mutual aid groups, and low-carbon household groups are examples. However, we also need to watch out for disinformation and misinformation on the Internet.

Systematic exposure to different ways of living from a young age can help simpler lifestyles emerge and can build networks of peer and family support over time. Social skills and environmental knowledge can be built within K-12 education and through local community institutions including religious organizations, clubs, unions, and environmental stewardship groups. College is also a great time for young people to learn to think critically about their values and what is important to them, as well as to develop peer support networks of others with similar goals. Employers can allow and encourage workers to spend time in service to the local community. This builds bonds among employees and between employees and community members, as well as improving a company's image and gaining it local goodwill. Health care workers, counselors, and mental health professionals can emphasize community-building and service activities as part of the solution to individual problems such as depression. They can also think more intersectionally about the ways in which environmental conditions contribute to health outcomes and work collaboratively with urban planners and community activists. Radical materialism has arisen for some good reasons, among them the lack of community and meaning in many people's lives and the lack of a social safety net. Positive alternatives and policy changes are needed to build back those things.

Take Advantage of Crises

Crises can help leverage behavior change. Sudden public consciousness of the crisis can motivate widespread adoption of new ways of living. The Second World War is often held up as an example of a time when North American society rapidly mobilized to confront an external enemy, with people changing their behavior dramatically so as to conserve resources at home that could be used for the military. Virtually overnight the US converted entire economic sectors such as the motor vehicle industry to war production. Perhaps the Covid-19 pandemic, the Black Lives Matter movement, the climate crisis, or future consciousness-raising events can have similar impacts.

Political leaders can exploit crises for bad as well as good. Populist politicians in recent years have sought to leverage perceived crises such as an influx of immigrants into political power for themselves. Proactive leadership is needed so that communities can prepare constructive responses for economic crises (for example, recessions, plant closures, or trade wars) or environmental crises (heat waves, drought, flooding, or hurricanes) when they arise and head off factions that would exploit the crisis for social division or narrow personal gain. We need to be anticipating crises and reimagining new, alternative visions for the future.

For example, one predictable future crisis in many parts of the world concerns water. Many communities are overdrafting groundwater and depleting river flows, at the same time that climates are often getting dryer and mountain snowpacks are diminishing. Cape Town, South Africa, is one such city. At least five years beforehand, officials predicted that "day zero," when the municipal water supply would need to be shut off, would arrive in early 2018. Politics, poor planning, and institutional dysfunction had led to this situation,[20] but as the identified date approached, leaders initiated a dramatic "day zero" messaging campaign, which combined with conservation measures and price increases led the residents to cut their water use by more than half. That change plus timely rains meant that a water shutoff was averted. A coordinated set of responses to an environmental crisis had produced massive behavior change.

CONCLUSION

Many urban sustainability and GHG reduction programs have ignored personal consumption and lifestyle questions. But radical sustainability and community reimagining require that all of us walk the walk, even as we advocate for bigger-picture structural change. It is impossible to deal fundamentally with problems such as global warming, poverty, racial injustice, and social inequality otherwise. An ethic of voluntary simplicity, a new focus on service and giving back to society, a sharing economy, adoption of plant-based diets, education and reproductive freedom for women, "families of choice" as an alternative to large biological families, and an emphasis on caring for each other and the world over acquiring more material possessions all can help. Seeing ourselves as agents of change empowers us to act and demand more systematic change.

Cities and other levels of government can use behavior change theory to identify strategies to help individuals reduce their ecological footprints. These strategies can include a combination of regulation, economic incentives, social marketing campaigns, education around lifestyle choices, and steps to create networks of support for individuals. Leaders can use crises to motivate changes in values and behavior, and can prepare ahead of time to use future crises in this way. In taking these individual steps, we may be able to avert the worst effects of climate and sustainability crises, and we may be all be able to live better, happier, and healthier lives.

10

How Can Cities Better Support Human Development?

Every year—pandemics aside—billions of us enjoy the vitality of urban places. We love the walkable streets, the range of things to do, the cultural facilities, the opportunities they afford us, and all the diverse and interesting people. We might attend a political protest or go to a museum. We might meet friends in a restaurant or pub, hang out at the playground, find a beautiful park or garden when we least expect it, or come across an intriguing work of public art. We might happen upon an interesting cluster of shops and restaurants and spend far longer than we had expected exploring them. We might talk with our friends and neighbors.

We live, work, and play in cities. Cities, at their best, help us grow as people by exposing us to new ideas, diverse perspectives, different cultures, and opportunities for civic engagement. They expand our understanding of the world and give us many options to develop our talents and creativity. They educate us and help us learn to care for diverse others. Supporting human development in these ways is essential to have healthy social ecologies.

Each of us could tailor our own vision of a city that feels fun, livable, educational, and nurturing. But today's communities do not offer nearly as many of these experiences as they might. Their physical environments can seem monotonous, boring, and traffic-clogged. Their social systems can be stressful or unjust. Many people are struggling just to get by. Governments often lack the funds to provide needed services to their residents or to enhance civic amenities. How could the form, institutions, and services of communities instead help people and social ecologies fulfill their potential? How could they do this for all groups within society, taking into account different cultural and individual needs and addressing past, present, and future injustice?

SOME HISTORICAL PERSPECTIVE

How cities can help develop human potential is a question that would never have been asked at most times in the past. Through much of history individuals lived in small villages that, though they may have had tight community bonds, had limited opportunities for personal growth and fulfillment. This was especially true for women, people of color, and those without social status and means. This world was prescribed and roles were limited and restricted. They did not offer people the full range of opportunities that we now know are essential to equitable human development.

As cities grew rapidly during the Industrial Revolution, life for many became downright grim. People worked ten or twelve hours a day for six or seven days a week in newly created factories, often for little pay and in unsafe and environmentally hazardous conditions. In the "City of Dreadful Night," as planning historian Sir Peter Hall termed the nineteenth-century industrial city, whole families lived in single rooms of tenement houses without running water or indoor bathrooms.[1] Coal smoke filled the air, and diseases such as cholera and tuberculosis were rampant. The situation was not much better in rural areas, where food was often scarce and many households depended on the mercy of large landowners. The lives of much of the population were dominated by the struggle to make a living and maintain health.[2] Unfortunately, this is still the case for billions of people around the world, particularly in the Global South. As we talk about reimagining sustainable cities, a first priority will be making sure that cities provide the basic services and amenities that billions of people still lack.

As countries industrialized, social movements pushed for human rights, humane working conditions, decent pay, access to education, equality for women and people of color, leisure time, and democratic participation in governance. Brave people the world over organized, struck, and at times paid with their lives to achieve such conditions, a fight that still continues. Basic elements of life that we take for granted today were hard won. One of our favorite bumper stickers reads, "The Labor Movement: The Folks Who Brought You the Weekend." These struggles to continue as we move toward a more equitable and sustainable world.

The urban planning profession, formally organized in Great Britain and the US during the first decade of the 1900s, helped lead efforts to establish park and trail systems, libraries, community centers, museums, senior centers, and educational institutions serving people of all ages. Wealthy individuals endowed theaters, dance troupes, orchestras, and private schools and colleges. Recreational opportunities became more common. However, many of these facilities were designed by and for white, wealthy men and were often exclusionary. They did not celebrate diversity or inclusion.

By midway through the twentieth century, social safety net programs in many countries provided health care, housing, disability and retirement income, and unemployment insurance, thus removing fear of destitution from the lives of millions. The spread of building codes led to much-improved and safer housing. Household appliances reduced labor for women. Women and people of color fought for their equal rights. Growing affluence went hand in hand with rising consciousness about human rights, environmental quality, social and racial justice, and public health. Opportunities for recreation, entertainment, and individual self-fulfillment expanded.

During the postmodern era, beginning approximately in the 1970s, urban culture became more creative and playful. Architecture moved to embrace color, variety, and creative juxtaposition of cultural forms. Public art became more whimsical and interactive, enlivening public spaces. Music and food from many cultures spread worldwide. Food trucks and other informal or pop-up uses of public space grew more common, although informal economies had long made use of streets worldwide, especially in the developing world.

Psychological study also became more focused on encouraging human development. Rather than focusing on mental illness and how to get rid of it—the emphasis of Freud and other psychological pioneers during the first half of the 1900s—humanistic psychology in the last third of the century sought to identify and promote human potential. After the turn of the millennium the field of positive psychology sought to learn lessons from happy, high-functioning individuals. The new discipline of happiness studies researched conditions leading to well-being. Starting in 2012, organizations affiliated with the United Nations began producing a *World Happiness Report,* based on self-reported individual-level data from many countries, which in most years has determined that Norway,

Denmark, or Finland is the world's happiest country. Sweden is usually also in the top ten though its ratings are hampered by growing pains associated with immigration. The United States and Great Britain are not among the top ten for many reasons, including lower social support, lower life expectancy, less freedom to make life choices, less generosity, and more perceptions of corruption.[3]

The process of planning and designing cities that promote happiness, quality of life, and fulfillment of human potential is still in its early stages. These goals have not been priorities, mainly because many urban jurisdictions are focused on more basic needs such as trying to create or maintain infrastructure and basic services. Neoliberal values of rapid economic development, deregulation, and privatization have dominated many societies so that funding creative, high-quality public services is difficult and is deprioritized. At the national level in the US, enormous spending has gone instead to the military, tax cuts for the wealthy, and subsidies for industries such as fossil fuels and corporate agriculture. We have not made the promotion of human development from birth to death a national priority.

Entrenched inequality is another reason why urban societies do not focus more directly on human development. Inequality begets greater inequality as elites seek to expand their power and wealth by capturing political and economic systems and keeping poorer classes disempowered. Many workers still lack living wages, good working conditions, decent vacation time (still a pitiful two weeks for many US jobs), paid parental leave, affordable childcare and eldercare, and health and retirement benefits. Such things are preconditions for individuals to be able to take advantage of educational and personal development opportunities. When people are so busy just trying to make ends meet, they do not have time to focus on finding joy in their lives or reaching their full human potential.

In terms of the built environment, a major problem is the scale at which communities are constructed and who does this. Place making is often undertaken by development corporations operating at very large scales, driven primarily by values of profit and efficiency. The prototype for production home building was Levittown on Long Island, where the company owned by Abraham Levitt and his sons built more than seventeen thousand virtually identical tract homes between 1947 and 1951. Soon other developers worldwide were using mass-production techniques. The cost of mass-produced housing may be relatively low, but at the expense of local variety, creativity, culture, history, and quality.

"Value engineering" is another version of this ethic applied to large public and private-sector building projects. Within this process every project element is reviewed to see if it is really necessary, and elements such as landscape design, high-quality materials, or high degrees of energy efficiency that have low economic return on the front end are often cut. Value engineering became mandatory for US government projects in 1996 and plagues private sector projects as well. The resulting projects often meet the economic bottom line but miss the mark in terms of creating beautiful spaces, public amenities, and energy self-sufficiency. The value that is being engineered is not about maximizing happiness, health, walkability, or long-term ecological benefit but about minimizing the front-end economic costs.

Twentieth-century building standards and codes likewise make it difficult to create innovative, sustainable, green, and equitable communities that support human development in new ways. Local officials often believe that streets have to be a certain width because of standards codified by state agencies or found within the "Green Book" produced by the American Association of State Highway Transportation Officials. These street widths are not designed not to enhance pedestrian safety or walkability but to maximize the efficiency of motor vehicle travel. Zoning codes often require excessive amounts of parking and prevent mixed-use development. Many regulations have good reasons behind them, but others are flawed and are based on outdated or biased assumptions. Twentieth-century codes often serve to lock in BAU, promote inequity, and lock out creativity and alternative visions for more sustainable place making.

The difficulties faced by anyone trying to create new types of communities are illustrated by Village Homes, a 1970s eco-community in Davis, California. This neighborhood is designed around greenways, edible landscapes, and community gardens. Homes use passive solar architecture (orienting structures to take advantage of the sun's warmth for heating) and are highly energy-efficient. Pedestrian paths run everywhere. Children can wander through the landscape with few barriers—and with little danger from cars.

However, creating this ecological neighborhood was extremely difficult, a labor of love on the part of two young, nonprofessional developers, Michael and Judy Corbett. The city planning staff initially denied permits, and plans were approved only because the city council at the time had been taken over by students, who overrode the staff's decision. The county public health department wouldn't approve gray water sys-

TABLE 10 STRATEGIES FOR CITIES TO SUPPORT HUMAN DEVELOPMENT

Strategy	Description
Physical Form	
Ensure varied, human-scale buildings and land uses.	Revise plans, codes, and design guidelines to add interest, variety, and sense of identity to built landscapes.
Require well-connected streets and paths in new development.	Make the city more accessible to everyone of any age; give people more varied ways to get around.
Create green, safe, and welcoming public spaces throughout the city.	Give everyone access to safe, restorative, and recreational out-of-door spaces; promote understanding of nature; reduce noise; use water as a design element.
Add art, music, and color to the urban environment.	Make cities more interesting and interactive; add whimsy, culture, creativity, and playfulness.
Make the city friendly to children, women, and the elderly.	In addition to humanizing urban spaces generally, tailor spaces to specific age and population groups.
Schedule public events, concerts, and shows.	Activate urban spaces through programming that provides a range of opportunities for enjoyment, interaction, and learning.
Public Services	
Improve educational opportunities.	Provide equally good, free public schooling for all through the university level; supply many types of continuing education and enrichment learning for all ages.
Support families and children.	Provide universal preschool, after-school activities for children, parental leave, and eldercare.
Ensure decent and meaningful work with opportunities to learn and collaborate.	Require better worker pay and benefits as well as engagement in managing workplaces.
Involve the public in design and planning processes.	Develop a wider range of civic involvement opportunities; require developers to undertake such engagement.

tems because they had never seen them before. The fire department wasn't happy about the narrow streets. No bank would provide loans because the proposed neighborhood was too different from conventional subdivisions. Only after the Corbetts rewrote their proposal to make the project look conventional would one credit union underwrite an initial phase. Though Village Homes has been very successful financially—and is seen worldwide as a leading example of sustainable residential design—no other developer has emulated it. The experience of the Village Homes highlights the difficulties of creating sustainable places that don't maximize economic return or meet existing norms. When we reimagine sustainable communities, we also need to reimagine the rules, financing, and norms that dictate and govern what we are allowed to build.

So what urban planning, design, and policy tools might help create cities that avoid such obstacles and better support human development, happiness, equity, and quality of life? Many different answers could be given, but following are suggestions in two main areas: physical form and public services, policies, and programs (table 10). Actions are of course interconnected. Foundational strategies outlined in previous chapters to reduce inequality are an essential starting point, as is universal access to food, clean water, shelter, childcare, and health care.

PHYSICAL FORM STRATEGIES
Ensure Varied, Human-Scale Buildings and Land Uses

As emphasized in chapter 7, more human-scaled and mixed-use urban form can add vitality to cities and can get people out of their cars. Where today large districts are monotonous repetitions of the same land use or building form—hundreds of single-family homes or apartment buildings here; large, boxy office buildings or sprawling, asphalt-covered malls there—reimagined cities of the future can provide many types of activities, services, and community amenities within walking distance. A long line of urban design theorists, most notably Lewis Mumford, Jane Jacobs, Kevin Lynch, Chris Alexander, Randy Hester, Nan Ellin, and Emily Talen, have argued for fine-grained, diverse urban form that is not dictated by the motor vehicle or large-scale corporate developers. Instead, it is dictated by people's needs. What are the services and amenities that every neighborhood needs?

One of the twentieth century's now almost forgotten intellectual titans, Mumford was enamored of small-scale medieval villages with tightly wound, organic street patterns and hoped to see a similar sort of organic unity within twentieth-century urban regions.[4] He warned against modern society's obsession with technology, rapid growth, and overproduction of low-quality goods, which he saw as leading toward the doomed "megalopolis." Jacobs loved the small-scale form and vitality of New York's Greenwich Village and later Toronto's Annex neighborhood, both of which she lived in.[5] She vigorously opposed large-scale twentieth-century urban renewal and the idea that supposedly objective white male experts would know best how to plan cities. She saw the layered richness in her neighborhood and thought the best planners were the citizens.

In a more academic vein both Lynch and Alexander distilled "good city form" values from millennia of history and argued for urban design principles aimed at creating a fine-grained, diverse, well-connected urban fabric.[6] Lynch's 1981 book *Good City Form* and the 1977 volume put together by Alexander and others called *A Pattern Language* were bibles to a generation of late twentieth-century urban design students. Hester emphasized grassroots community involvement to identify "sacred places" and build upon local community strengths.[7] He and his partner Marcia McNally guided groups of local residents around their own community, identifying and mapping the places that were important to them. They drew small red hearts on maps to identify sacred places. Such locations usually weren't connected with religious institutions or large public plazas but instead represented humble locations such as a local post office or waterfront where people met, socialized, and built a meaningful web of connection. These were places that had cultural significance and meaning—and that could be preserved and prioritized as the city developed.

More recently, Nan Ellin concisely skewered the contradictions of large-scale, corporate postmodern design, which has often created inauthentic mishmashes of pseudohistoric urban form aimed at upscale consumption.[8] She has advocated instead a diverse weave of uses, styles, spaces, and functions across cities aimed at allowing multicultural populations to live, work, and enjoy urban environments.[9] Talen has been closely associated with the New Urbanism and emphasizes institutional changes to develop new design and planning codes that can create a diversity of forms and spaces across the city.[10]

Such thinkers help us understand a near consensus that has emerged among urban designers in the early twenty-first century. In contrast to the large-scale, bland, minimalist urban landscapes envisioned by twentieth-century modernists, oriented around values of efficiency and mobility, this view sees the future city as having a diverse, fine-grained, human-scale, and pedestrian-oriented form, prioritizing human and ecological welfare. Such communities might mix some single-family homes on small lots with duplexes, townhouses, small apartment buildings, and taller buildings. Neighborhood centers would contain stores and workplaces. Green spaces, playgrounds, public plazas, and pedestrian and bicycle facilities would abound. Combined with good schools and services, this mix would provide the density and variety needed to generate urban vitality and a wide range of opportunities for human development at all stages of people's lives.

What will it take to achieve such form? Good, community informed plans can help. Plans aiming at creating "urban villages" have proliferated in recent decades. Seattle developed a general plan with an urban village theme in the 1990s, and San Diego's General Plan organized itself around a "City of Villages" strategy in 2008. Phoenix, San Jose, Fort Worth, and Belfast in Northern Ireland have recently embarked on urban village initiatives.

Changes in codes are essential as well. At times zoning codes may need to be relaxed to allow a greater variety of building forms and land uses and to prepare for climate resilience in new ways. Or else form-based codes, discussed earlier, need to be adopted instead. At other times new requirements may need to be added to codes to prevent the sorts of large-scale corporate development that otherwise dominate. For example, cities can limit maximum building footprint size, require a mix of building and unit types, and prohibit drive-through fast-food franchises. In older cities, historic preservation regulations may be critical to protect neighborhood character from the influx of new developer-led construction and to prevent rampant gentrification. Institutional changes may also be needed to allow for more creative planning and equitable placemaking. Since US municipalities get much of their budget from fees and taxes on newly built commercial development and upscale housing at the urban fringe, state or national government may need to institute new tax-sharing policies to eliminate this incentive for continued sprawl development and the resulting regional inequity and habitat fragmentation.

Require Well-Connected Streets and Paths

Street patterns establish the armature of cities and towns. They determine the size and configuration of blocks, the walkability of neighborhoods, and often the scale at which development takes place. Suburban neighborhoods with looping roads and cul-de-sacs are almost by definition unwalkable, since very few streets connect through to others.

Improving the connectivity of urban street patterns is a main way to encourage human use. Connected street and path networks are relatively easy for cities to require within new development. However, retrofitting the existing urban fabric is more difficult. Sometimes new paths and roads can be cut through vacant parcels, or new pedestrian and bicycle paths can be added along utility easements or old railroad rights-of-way. Overly wide streets that have become barriers for pedestrians and cyclists can be redesigned with new bike lanes, sidewalks, and planting strips. Pedestrian promenades around parks or along waterfronts are another way to add pedestrian circulation, improving human use of the urban environment. Or more radically, there may be times when it is necessary to take over the streets entirely for play and green space, putting people, not cars, first.

The speed of vehicles in a city is a central quality-of-life issue. Above about 40 mph, vehicle noise, fuel use, and GHG emissions increase. If vehicles go above 30 mph, pedestrian safety and comfort sharply decline. Directly contrary to the transportation planning goals of the 1950s, a traffic-calming movement arose in northern Europe beginning in the 1960s that has led to street design changes in many cities worldwide. New street designs can allow both slow and steady motor vehicle traffic and an attractive, pedestrian-friendly streetscape. The toolbox of strategies for slow streets includes corner bulb-outs to increase pedestrian safety; speed tables (broad, raised crosswalks); planted medians and islands within the street; narrowed lanes; addition of bike lanes; addition of parking to buffer sidewalks from motor vehicle traffic; and land use strategies to line streets with actively used buildings rather than parking lots or large, empty lawns.

Create Green, Safe, and Welcoming Public Spaces throughout the City

During the twentieth century, motor vehicles took over much public space, and civic life in many parts of the world retreated into private or quasi-private spaces such as malls, gated communities, privatized parks, and shopping centers. Now cities need to prioritize, invest in, and rebuild the public realm, guided by equity and planning for climate resilience. Getting housing within all parts of the city, including previously lifeless business districts, is a prerequisite in order to populate public space. Having people on the street at all hours makes neighborhoods more interesting, safe, and fun. Making cities fun, safe, and playful for children is essential. "Eyes on the street" is a time-tested maxim for improving public safety. Spaces for performers and sidewalk vendors helps create local attractions. Sidewalk seating for restaurants is a way both to bring people into the public realm visible to others and to offer them a place where they can enjoy the company of their neighbors while eating a meal.

Public spaces themselves can be designed in many different ways. What is most important is a mix of spaces, large and small, formal and informal, so that everyone feels welcome in all types of communities. A green space and a public plaza should be within a ten-minute walk of every dwelling. If well designed, each individual park or square can have many different outdoor "rooms" within it. As William H. Whyte famously documented in his book and film *The Social Life of Small Urban Spaces,* good design makes a big difference in how much people use and enjoy public spaces.[11] Places with good seating, sun, a variety of public and semiprivate spaces, opportunities for people-watching, activities for children, access to nature, and water are likely to succeed and add to the vitality of the community. Cities can require developers to add public spaces, including play areas, gardens, and landscaped plazas, within their projects, sometimes giving them a bonus in terms of increased size or height of development.

One important design goal is to make outdoor spaces restorative, which according to psychologists Rachel and Steven Kaplan includes adding elements that are "fascinating," such as water, art, creative and native plant design, or animals. These draw people's attention away from stressors in their lives. Other important design goals include elements that give people a sense of "being away" from the world for a few moments, and elements such as grottos, groves, or views that provide a sense of depth.[12]

Water features particularly draw people and add interest to cities. One of the reasons that Paris is one of the world's most loved cities is its many fountains, not to mention the River Seine with its bordering promenades and, in the summer, an artificial beach. Traditionally many cities have kept people away from water features, strictly forbidding putting any part of one's body in the water because of safety and hygiene issues. But why not make sure that the water is swimmable and let people interact with water? Cities such as Scarborough, Ontario, Boston, Massachusetts, and Portland, Maine, allow children to wade in shallow water features while adults picnic or sit at outdoor cafés and watch them. Philadelphia's City Hall has a water spray park that is used by kids and adults to cool down in the summer. In the winter, the facility turns into an ice-skating rink.

Excessive noise can harm hearing, disrupt sleep, and lead to stress, anger, high blood pressure, and many other ills. Noise has many causes, among them motor vehicle traffic, airplanes, construction, industrial activities, and loud human voices. All of these can potentially be addressed. We may need to require particular types of vehicles such as scooters, two-cyl-

inder motorcycles, and heavy trucks to have better mufflers (called silenc-
ers in Britain). Moving to electric vehicles—necessary anyway to reduce
GHG emissions—will virtually eliminate engine noise. Lowering the volume and speed of traffic helps reduce noise; this can be done through slow
street design, as discussed above. Street design that avoids stop signs and
stoplights in favor of other traffic-slowing techniques such as roundabouts
can reduce the noise, pollution, and GHG emissions produced when vehicles start up from a stop.

For their part, airplanes have become substantially quieter since the
middle of the twentieth century as jet engines have been replaced by
turbofans. Further improvements may be possible. Many cities require
reduced throttle when commercial planes are flying low over urban
neighborhoods or have required that flights be routed over less populated areas. Regulations about construction and industrial noise are
also possible, for example prohibiting the start of such activities before
8:00 a.m. or requiring noise abatement protocols.

Urban residents benefit greatly from outdoor activity options. Walking and jogging paths, bike routes, pools, exercise equipment, basketball and tennis courts, and playing fields are already priorities in many
communities. Newer types of facilities include skateparks, volleyball
courts, skating rinks (in season), and paddleball courts, as well as more
meditative spaces such as labyrinths and memorial sites. Bodies of water
where people can go kayaking, sailing, or paddleboarding are also a
good way to connect city dwellers to the natural environment. Examples of playful, interactive urban park landscapes include the High Line
in New York City (an old elevated railway converted into a three-mile
linear green space) and Parc La Villette in Paris (the former site of a
slaughterhouse transformed into a network of whimsical spaces).

Subdivision codes can require a range of shared outdoor spaces
within new development. Davis, for example, requires that new subdivisions include greenways connecting to the existing citywide greenway
system. Restoring or adding biophilic elements to cities, as discussed
earlier, dovetails with this objective.

The public is often concerned about the safety of urban spaces,
although ironically some suburban environments may be more unsafe
because of a higher frequency of traffic accidents and fewer eyes on public spaces. Safety is not a simple problem and varies enormously across
contexts. Good urban design can help, for example arranging and designing buildings so as to maximize eyes on the street and balancing land
uses to as to keep people present within streets and public spaces at all

hours. But more fundamental policy changes will also be needed to reduce poverty, meet human needs, and provide opportunity to all members of society. A society that cares for its members in these ways is likely to be far less dangerous. In the US, gun control policies and reforms to reduce police militarization and violence will also be essential.

Add Art, Music, and Color

Nothing embodies a community's commitment to the arts better than public art. Playful, colorful, unexpected forms within civic spaces help humanize these areas and attract people. Not for nothing have cities worldwide focused on this strategy in recent decades in order to improve urban vitality.

Public art has evolved dramatically over the past century and will need to continue to evolve to be more inclusive and meet equity goals. Nineteenth-century white-men-on-horses statuary gave way to modernist sculptures in the twentieth, though highly abstract versions were often ignored by the public and derided as "plop art." Though they may have had color and creative form, works composed of steel girders or other abstract elements held little interest for passers-by. The postmodern era has seen emphasis on whimsical, interactive installations that often draw on local culture or ecology and entice people into engagement. These installations have been more successful at making urban places fun and at times educational. Digital installations that allow people to interact with the artwork with their mobile devices also have a lot of potential to engage and educate residents.

One iconic interactional projects was "Cows on Parade." The first of these temporary installations of fiberglass animals along sidewalks occurred in Zurich in 1998, followed by a large-scale "cows on parade" exhibit in Chicago in the summer of 1999. The city asked local artists to decorate two thousand full-size fiberglass cows, each choosing a distinctive style and concept. These colorful animals were then installed along city streets and within parks. Tourists and locals alike loved them. Kids climbed on them and adults photographed each other next to them. At the end of the specified period Chicago auctioned the cows off for charity. In the early 2000s many other cities followed suit. Washington, D.C., installed elephants and donkeys. Jerusalem chose lions, and Dubai camels. Portland, Maine, featured lighthouses. In a few cities some of these works of art became permanent, adding a colorful and whimsical element to the streetscape.

Interactive art beckons to both kids and adults and is a natural conversation-opener for strangers. Ideally people can climb on the art, move pieces of it, walk through it, ride it, see themselves reflected in it, or engage with it in some other way. The *Cloud Gate* sculpture in Chicago, by Anish Kapoor, is a good example. A polished, reflective, curving bean of polished steel thirty-three feet high, designed to look like a gigantic drop of mercury, it is a magnet for pedestrians who can see themselves reflected with various interesting distortions.

Cities can require developers to include public art within all large development projects and encourage hiring diverse local artists. The US municipalities of Philadelphia, Los Angeles, Atlanta, Portland, Oregon, and Tampa have all put in place "Percent for Art" codes, under which a certain percentage of the cost of any municipal improvement project must be devoted to art. The usual figure is 1 percent, although San Francisco and San Diego have required 2 percent. Usually a public arts council makes the decision about how such funds will be spent.

A final frontier is to encourage the design of urban buildings to be diverse, playful, and colorful. Such initiatives are typically embodied within urban design or architectural guidelines for downtown districts and aim to produce a lively, interesting landscape in contrast to the often-monotone gray world of mid-twentieth-century modern development. Cities can use design review committees to likewise encourage lively architecture. This type of effort is still in its infancy, however, with many new buildings continuing to exhibit bland and boring design.

Make the City Friendly to Children, Women, the Differently Abled, and the Elderly

Children are an indicator species for healthy cities, and when they are cared for so are the rest of us. If they abound in public places and on public transit, then the overall community is probably a good place to live, and people are probably growing and thriving. But how do we make cities kid-friendly? One prerequisite, of course, is that lead and other environmental hazards in older cities be cleaned up before young people use urban spaces. Beyond that, addressing gun violence, improving public safety, and decreasing vehicle speeds are good starting places. Adding vegetation, water, gardens, natural spaces, and public art help as well.

Specific play areas for children of different ages are important. Infants and toddlers need different types of facilities than older kids. Sandboxes and ground-oriented playground equipment are best for the former.

Climbing structures, swings, magical spaces, and facilities that improve strength and balance are suited to the latter. Wading pools, surprise fountains, and other sorts of water features attract children of all ages. Teenagers are difficult to design for but can benefit from safe and secluded hang-out spaces as well as more active exercise facilities, such as skateboard parks, bike tracks, basketball courts, pools, and running tracks. Programming and educational opportunities for teenagers are important, particularly after school and during the summer and other school breaks.

Other design elements can make it easier for parents to bring kids into urban spaces. Benches and shade near play areas are important for grown-ups needing to wait while kids play. Cafes with outdoor seating help as well. Restroom and infant changing areas are crucial. Dog spaces are important for many urban residents, for whom walking their pet one or more times a day is part of the daily routine. Those spaces may need to be separate from playing fields and other park spaces, since dogs need to be able to run off leash in ways that might disrupt human activities.

We also need to care for and plan for aging populations and differently abled people. In many cases the same amenities, such as curb cuts that make it easier to push a baby stroller also make it easier for wheelchair users or people who need other mobility assistive technologies to navigate the city. Benches and other resting places are useful for the elderly. The idea of universal design is critical to making sure our cities are equitable and accessible. Everyone needs access to the amenities that the city provides. In cases where older people or people who are differently abled need more assistance, we need to develop programs to meet their needs.

Schedule Public Events, Concerts, and Shows

Creating wonderful, well-used public spaces isn't just a question of a well-designed space in the right location. The most successful cities actively program key spaces. Events can include concerts (noon, after work, and evenings are prime times), dance performances, theater, comedy, festivals, markets, or even circus events. Events linked to local culture and history are particularly desirable. For example, if there is a substantial Ukrainian, Vietnamese, Brazilian, or Somali population in the area, musical or dance performances from those ethnic traditions can be recruited. Having a regular and well-publicized schedule of free, culturally appropriate activities helps activate a place. For example,

Portland, Oregon schedules events ranging from dance performances to meditation in its Pioneer Courthouse Square downtown, with an online calendar providing information to the public. Conversely, pop-up events can add interest and intrigue for those simply passing by. Magicians, jugglers, and street musicians can be encouraged to perform at all hours in public spaces and be paid for their labor. Either approach can give people a reason to spend time in the public realm.

Programmed activities in public spaces should of course be open to the public. There is a recent trend to cordon off parks and plazas for special events with hefty admissions charges, often to raise money for the parks themselves. This is one result of the neoliberal ideology that everything should pay for itself. However, many if not most public goods don't pay for themselves. This is why we have government—to ensure clean air, good schools, parks, public health, police and fire protection, and many other social goods. In addition, many benefits of public spaces and green spaces cannot be quantified but are certainly very valuable. For instance, green space can reduce stormwater runoff, contribute to reduction in urban heat island effects, and be important for public health. Antitax crusades ignore this basic reality. Professionals and political leaders need to help the public understand the necessity of a strong public sector and investment in public amenities that provide public benefits.

PUBLIC SERVICES STRATEGIES
Improve Educational Opportunities

Nothing is more central to human development than education. Whether it is continuing education for seniors, universities, workforce training for young adults, K-12 school systems, or preschool for young children, high-quality education available to all residents (not just those who can pay to live in expensive communities or attend private schools) is a cornerstone of high-functioning social ecologies. Across societies, educational expansion has been shown to improve individual health and economic security.[13] Education is also highly correlated with democracy and should be free or nearly so.[14] The idea that Americans should mortgage their futures to pay for college or graduate school debt is insane.

Well-funded, universal, free preschool programs can play an enormous role in determining a child's life chances, especially for children toward the lower end of the socioeconomic spectrum.[15] Providing this to every one of the twenty million children under age five in the US—while paying well-trained childcare workers good salaries—would cost some $200 billion a year, assuming average costs of $10,000 per child (an early 2010s proposal by the Obama administration that had greater state contributions came in at $79 billion). This might seem an impossible amount. Yet the country could easily afford it. For example, Americans—primarily the wealthy—underpay their taxes by $381 billion annually.[16] One might think that Congress would be anxious to recapture this enormous amount by adequately funding the Internal Revenue Service to catch tax cheats. But Congress has cut the agency's funding routinely since 2010. The number of returns audited from tax-payers earning more than $1 million has fallen 72 percent during that time, and the number of Americans who do not even bother filing tax returns has risen 20 percent. With such low-hanging fruit—not to mention excessive military spending and subsidies for fossil fuel companies and big agribusiness—availability of funds is clearly not the reason the US fails to support its educational systems.

Support Families and Children

Caregiving cities that support human development must be designed particularly with women, children, and the elderly in mind. According to geographer Leslie Kern, "A feminist city must be one where barriers—physical and social—are dismantled, where all bodies are welcome and accommodated. A feminist city must be care-centered, not because women should remain largely responsible for care work, but because the city has the potential to spread care work more evenly. A feminist city must look to the creative tools that women have always used to support one another and find ways to build that support into the very fabric of the urban world."[17]

What would this mean? Perhaps most importantly, a range of social programs can make it easy for parents and children to thrive in communities. European countries offer generous paid parental family leave and inexpensive childcare. Cities might sponsor after-school programs and camps so that parents with jobs are not caught between their employment and their need to take care of kids in the afternoon and on days off and summer breaks. Employers can be required to offer flexible work schedules and advance scheduling to meet parents' needs. Public transit can be made easier, safer, and cheaper for children and families.

Design interventions are important as well. Simple devices such as benches, shade structures, covered bus stops, and ramps to facilitate pushing strollers or bicycles up steps can help. Well-maintained public restrooms and diaper-changing facilities (in both men's and women's rooms as well as all-gender rooms); cafes and water fountains in parks; childcare centers near places of employment; and interactive play equipment throughout the urban environment are all ways to make cities family-friendly.

Ensure Decent and Meaningful Work with Opportunities to Learn

Most of us spend the largest chunk of our time (except for sleeping) at work. If our jobs are boring, repetitive, underpaid, and not respected by society, it is hard to see how our overall lives can be as happy and fulfilled as they might be. We've mentioned the first step toward addressing this problem many times already—radically raising low incomes within societies while capping or heavily taxing high incomes. There's no reason why a wealthy person contributing little if anything to society should make enormous amounts off investments while a schoolteacher or childcare worker performing an extremely important social function is barely able to pay rent. Dramatically raising wages across the bottom half of the income spectrum and capping them at the top end, through regulation or tax policy, are essential for a healthier, happier, and more equitable society.

Starting pay for McDonald's workers in Denmark is about $22 an hour (almost three times as high as in much of the US), with six weeks paid vacation annually, paid sick leave, paid maternity leave, life insurance, and a pension plan (plus the government covers health insurance).[18] Why shouldn't low-end American workers be treated in such a decent fashion? During Covid-19, we saw workplace inequities affect such workers in other ways as well, as many minimum-wage workers had to continue performing in-person essential services, often at great health risks, while white-collar workers were able to work from home. We saw what it meant to be "essential" and how dependent we are on the well-being of these workers.

Going further still, we can explore the question of how to restore dignity to manual labor. Decent working conditions are a starting point. In recent years Amazon has required that many workers take no more than one and a half minutes for restroom breaks. Such rules are utterly disrespectful to workers and disregard their humanity. In addition to better work rules and conditions, ample vacation time, sick leave, parental leave, and retirement benefits would help essential workers feel respected by society. Worker councils and other means of giving workers input into the running of businesses and organizations would help make everyone at all pay grades feel valued, like a full-fledged, valued, and contributing member of society.

Involve the Public in Planning and Design Processes

Improving quality of life in communities must be a collaborative process engaging all residents and stakeholders and seeking out those who have been traditionally excluded. Top-down initiatives are unlikely to meet the needs of all. Instead, they may antagonize some communities and divide people. Public involvement processes often take time and money. Officials need to listen to people's stories about their community and what they imagine for their children and grandchildren. Well-run public engagement processes are essential and usually well worth it. Professionals who specialize in community participation can run design workshops, charrettes, focus groups, and hands-on events in which local residents become directly involved in improving their neighborhoods.[19] People who are not professionals must be included in designing these processes (and paid for their work), since their voices are often ignored. The next chapter takes up this subject in greater detail.

CONCLUSION

In the twenty-first century, cities and towns can become far more supportive of human development and well-being. They can become more fun, vibrant, and restorative places in which to live, places where creativity and cooperation flourish while people learn from others different from themselves. Many urban design theorists agree on physical form elements of such a city—it would be more human-scaled, with a greater balance of land uses and building types as well as great public spaces that meet the needs of diverse populations. In a reimagined sustainable city, the public realm would be designed to be educational, welcoming to all types of people, and restorative in many places. Public programs, events, classes, family activities, nature, and facilities such as community gardens can help as well.

As with other topics, what's important is a package of policies, programs, and design strategies that reinforce each other and are intersectional. Questions to ask include "Are cities meeting the needs of all of their residents?" "Do we have a culture of care?" "Are there certain assumptions about who is entitled to 'use' the city that need to be challenged?" and "How do we ensure that people from all walks of life (ethnic, religious, cultural, racial, age, gender, income, and ability) are welcome?"

Structural barriers to addressing these questions are primarily political. Most societies have the money to meet human needs in these ways if they choose to. The problem is priorities and distribution. Will funds be allocated for human development or allowed to overflow the pockets of the wealthy? Will they be spent to improve social and environmental well-being or wasted on unnecessary military hardware? Political leaders will need to make the case for the former. Healthy societies depend on the individuals within them being empowered, creative, and fulfilled. Cities can support their residents' development in these directions. They can be places of joy and potential.

How Might We Have More Functional Democracy?

It's a scene many of us have witnessed time and again: a crowded city hall, dozens of angry citizens, a stoic city council, a couple of patient but beleaguered planners at a table up front. Members of the audience are filled with righteous anger. Listen to a couple of them and their concerns appear justified. Why weren't local residents notified earlier? Why didn't the architect design the project to be more in keeping with the neighborhood character? What about parking? What about traffic? What about concerns about neighborhood change?

But then advocates of the project speak. The approval process has already taken nearly a year, they say. They dropped notices around the neighborhood from the very beginning. They've answered the same questions numerous times. Where else is affordable housing going to go? The site is near public transit and stores. There is an enormous need for housing not just for minimum-wage employees but for teachers, nurses, and police officers who can no longer afford to live in the community.

The hearing grinds on hour after hour. It gets to be midnight, or later. Everyone wants to go home. No one is enjoying this process. People feel frustrated. They feel that they are invited to the process after decisions are made, that they are not being heard, that this process is not meaningful and not truly participatory.

Although the subject may change, this scene repeats itself again and again within local government. Public anger butts up against long, drawn-out processes and conflicting local priorities. Everyone gets burnt out. Is this the best we can hope for in terms of local democracy? A war of attrition between constituencies? Let's hope not! How could societies get away from incremental, parochial thinking and move toward more robust, intersectional, equitable, and sustainable policies that are informed by community concerns?

The question of how to have better-functioning democracy is particularly pressing in the current era. Populism, apathy, disillusionment, election tampering through social media, gerrymandering of legislative districts, corporate influence on politics, and self-interested leaders are just a few of the problems. Structural racism is increasingly acknowledged, and we need to do everything we can to address it. So how can decision-making happen in more open, creative, collaborative, participatory, and constructive ways, not just within local government, but at other scales and across scales as well?

THE SLOW EVOLUTION OF DEMOCRACY

As often when confronted with depressing situations, one can brighten the picture by realizing that things were even worse in the past. Efforts at participatory government are very recent, becoming widespread only during the past century or two. Systematic public engagement within urban and regional planning dates just to the last fifty years or so. The concept of proactively bringing disadvantaged groups into public processes was rarely considered before then. Kings, dictators, political elites, large landowners, industrial titans, and their hired professionals simply went ahead and did what they wanted. The results were usually anything but just.

Even after the Second World War, mammoth development initiatives in supposedly democratic countries proceeded with little public involvement. For example, British, Swedish, French, Danish, and Dutch governments built new towns outside or at the edge of existing cities with little public consultation. The British new town program constructed thirty new cities housing two million people with virtually no input from the public on how they should be designed. Although some of these places work reasonably well today as suburban centers, many others have been criticized as sterile, boring, inequitable, and motor vehicle oriented.[1]

When demands for public participation in urban planning arose in the late 1950s and 1960s, it came as a shock to many planners and public officials. Jane Jacobs was the archetypical example of the new breed of nonprofessional activist. A journalist living in New York's Greenwich

Village, she worked with fellow residents to organize protests against planning czar Robert Moses's efforts to run a freeway through her neighborhood. Jacobs and her peers secured favorable media stories and won support from local politicians, in the process saving Washington Square Park and getting motor vehicles removed from it. Her book *The Death and Life of Great American Cities,* published in 1961, remains probably the single most influential piece of urban planning literature from the twentieth century. In it, she calls on readers to see the value in traditional dense, mixed, small-scaled urban forms and to value and listen to the residents of such neighborhoods. Her vision is the opposite of top-down, expert-driven, and exclusionary planning

Also starting in the 1960s, civil rights activists demanded more community control over urban redevelopment projects in order to redress social and racial injustices. Environmentalists began demanding environmental review of projects. Historic preservationists launched campaigns to save old buildings and neighborhoods. Ethnic groups, civil rights groups, feminist leaders, disability activists, and gay rights advocates demanded that their perspectives be considered. European grassroots leaders often focused on the growing impacts of motor vehicles. In a flowering of civil society, these activists formed a wide range of new organizations to gain a greater "right to the city" and make sure that their concerns were heard by decision makers.

Sherry Arnstein's classic "ladder of citizen participation" article in 1969—still the most read article in the *Journal of the American Planning Association*—gave voice to the citizen backlash against exclusionary, top-down planning.[2] In her ladder diagram, Arnstein distinguished between minimal or tokenistic public involvement practices at the bottom rungs, which she called "manipulation," "therapy," "informing," and "placation," and more meaningful responses such as "partnership," "delegated power," and "citizen control" toward the top. This conceptualization flipped the traditional power structure on its head. Local stakeholders, not experts, should be in charge of projects, in her view. Around the same time Paul Davidoff and others called for "advocacy planning," in which planners would work proactively with underrepresented constituencies to help them gain power.[3] The assumption was that planners needed to work to undo past injustices. Such processes would be more democratic and would create policies that better met people's needs.

In the US, the Johnson administration's Community Action and Model Cities programs, both aimed at reducing poverty, mandated public involvement. Environmental review processes legally required public involvement beginning in 1969. A new generation of community organizers, including a young Barack Obama, worked with urban neighborhoods to help them exercise their voice in municipal decision-making. New philosophies of urban development emphasized the extent to which public engagement could lead to better decisions. Communicative planning theorists emphasized consensus-building processes between stakeholders. Neo-Marxists focused on the role of social movements in leveraging change from outside government. The ABCD approach championed by Kretzmann and McKnight argued for economic development based on working with local organizations and identifying local strengths. Concepts such as the "coproduction" of public services—a situation in which the public sector works together with NGOs to efficiently and contextually provide services to the public—gained popularity.[4]

Other countries also spread political power more broadly during the last third of the twentieth century. Sweden and Denmark promoted tenant involvement in the planning and management of housing developments, as did designers of affordable housing in Argentina.[5] New Zealand adopted broad citizen participation requirements for local government in the 1980s, as well as specific requirements that the Maori indigenous population be included and that sustainability policies be developed.[6] The Netherlands emphasized consensus-based processes in which representatives of business, government, and environmental organizations developed five generations of a National Environmental Policy Plan beginning in 1989.

New civic engagement techniques emerged during this period. Urban planners and designers experimented with focus groups, surveys, visioning exercises, walking tours, cognitive mapping, block exercises, 3D models, idea contests, scenario planning, computer-based gaming tools, and consensus-based stakeholder processes. Movements such as the New Urbanism promoted design charrettes—urban design-oriented community workshops—as a way to involve residents in place making. After the mid-1990s, email and social media made it possible to deliver materials to large numbers of residents quickly. Planners used games and simulations as a way to engage with the public online. Municipal websites and apps provided an opportunity to put data and documents where members of the public could easily download them, and conversely to collect data and opinion from members of the public. Tools such as story mapping—which can refer to multiple techniques to develop community narratives and link them to places and processes—are increasingly used to learn from people's experience living in particular communities.

Stories provide the intersectional lens to understand how people experience the cumulative impacts of urban and environmental conditions.

Inclusive, effective governance is now globally acknowledged to be an important sustainability goal. Sustainable Development Goal 16 ("Peace, Justice and Strong Institutions") sets a target to "ensure responsive, inclusive, participatory and representative decision-making at all levels." So how do we make it work better in practice? There seem to be two conflicting needs: rapidly addressing sustainability problems such as the climate crisis, racial and social inequity, and uneven development, and effectively involving large numbers of individuals and stakeholder groups in decision-making.

TWENTY-FIRST CENTURY OBSTACLES

NIMBYism occurs worldwide and is one main obstacle to rapid action for urban sustainability, especially for affordable housing, which is one of the main tools to address social and racial justice dimensions. Local opposition to projects is justified in some cases, for example when developers ignore community context, place unjust burdens on residents, or wish to demolish historic buildings that merit being saved. But in other instances NIMBYism simply displaces people's generalized anger onto a handy local object. Our current local decision-making frameworks often reflect racial or class prejudices that have caused and in turn are further aggravated by a fragmented local governance structure. When we make decisions at the local scale without thinking about the larger implications of these decisions, the result is that many public needs at the regional scale—for affordable housing, equitably distributed school funding, social service facilities, bike and pedestrian trails, and public spaces—go unfulfilled.

Allied with NIMBYism is the rise of political polarization—or, to be more specific, a right-wing populism that rejects immigration, social, racial, and religious diversity, the equality of women, climate science, and other phenomena that it sees as threats. Ground zero for such sentiments is the white working class, a group that feels threatened by diversity, culture change, and economic globalization. Ironically, such voters don't see that the right-wing politicians they support allow the wealthy to further exploit them.[7] Instead, right-wing leaders, media, and culture have led them to oppose progressive policies designed to create stronger social safety nets and address injustice, which would usually help them as well.

Polarization is fanned in part by misinformation and a distrust of science and expertise, which are criticized as "elite." To be constructively engaged as citizens, people from childhood on must have access to good information on contemporary topics and be surrounded by multiple points of view. They need to learn how to discern real information backed by evidence and reason from misinformation. This is particularly important given the vast amount of the latter on the Internet, and the fact that much of the US gets its information from Fox News, a channel that conservative media mogul Rupert Murdoch and Republican operative Roger Ailes launched in 1996 specifically to promote right-wing viewpoints. Although until 2017 Fox used the slogan "Fair and Balanced," it has been exactly the opposite—a propaganda channel for far-right viewpoints. Right-wing talk radio, enabled when the Reagan administration and its allies in 1987 scrapped the Fairness Doctrine requiring radio and TV stations to provide "reasonable, although not necessarily equal" opportunities for opposing political viewpoints, exerts a similar effect.[8] For more than three decades, talk radio and Fox News have exerted a profoundly negative influence on American democracy and have undercut hope of progress on many sustainability issues. In particular, they have minimized and questioned the climate crisis, often viewing it as a liberal hoax designed to promote more government intervention and justify spending.

Other information deficiencies exist as well. Because of the growing consolidation of media outlets, many communities have no local newspaper that might report on local politics and provide citizens with multiple points of view on local issues. Radio, TV, and cable systems are often owned by multinational corporations seeking only to maximize ad revenue. Local TV news reporting is often limited to sensational issues that gather the most attention. "If it bleeds, it leads" is the unwritten maxim of many local stations. Increasingly, younger people are also relying on social media for their information, much of which is not vetted or fact-checked.

Power imbalances continue to threaten democracy and good decision-making at all levels. Constituencies with a financial stake in local decision-making—developers, landowners, large businesses, construction companies, realtors, attorneys, and consultants—are usually overrepresented at public events and on city councils and commissions. Meanwhile, citizens are overloaded with the activities of daily life and have difficulty participating in public processes. They are asked to volunteer their time to participate in community decision-making. This is time that most people do not have to spare. Workdays are long. People are underpaid. Families, children, eldercare, and housework consume nonwork time. As a result it can be hard for people to follow local politics closely, or for planners to get people to turn out for civic events of any kind. The best-advertised public meetings may be attended by only a handful of citizens (often those with more time and resources than others). So while the intent may be to bring people to the table, the fact that the meeting is held on a school night without free childcare may be the reason that people do not show up, not that they do not care. If we want more meaningful participation in planning decisions, we will need to reimagine participation so that it is inclusive, meaningful, and even paid.

Given how intersectional the challenges we face are, another challenge is being able to understand the big picture and the way that decisions interact. Unfortunately, current educational systems don't nurture citizenship skills as much as needed. Recent official emphasis in many countries on the STEM disciplines—science, technology, engineering, and math—often comes at the expense of producing well-rounded citizens with skills of critical thinking, analysis, synthesis, cooperation, leadership, and empathy. An understanding of history, politics, culture, communication, and ethics is also critical for good and effective citizenship. Right-wing organizations have fought against such "humanistic" education, believing it a threat to traditional value systems. But sustainable societies need citizens who can understand the complexity and intersectionality of current issues, place them in historical context, understand scientific and economic models, and see through the flimsy promises and self-serving rhetoric and manipulation of many politicians and pundits.

A final challenge to local democracy is less dramatic but just as profound: the lack of public sector capacity and the fragmented and siloed nature of government. Many cities and towns just do not have sufficient staff or resources to address sustainability issues. The fact that governance of metropolitan areas is fragmented into dozens or hundreds of jurisdictions also works against development of collective solutions. Local officials plan for their communities without regard for the impacts of those decisions outside municipal boundaries. Communities compete for revenue and development instead of collaborating to plan for equitable, green, and just regions. The department of housing does not coordinate with the parks department. Conservative efforts to cut taxes in many countries have intentionally hampered government—aiming to "starve the beast," as Ronald Reagan put it—at a time when the public sector needs to increase its capacity to lead on topics ranging from global warming to social inequality. Limiting public sector capacity in this way has allowed the wealthy and corporations to expand their dominance while the public becomes increasingly beholden to the private sector or to public-private partnerships. These are inefficient and inequitable approaches to meeting public needs that must be reimagined.

STRUCTURAL CHANGES FOR MORE FUNCTIONAL DEMOCRACY

Many factors, in short, work against effective democracy, not least the outsized influence that corporations and the wealthy have over politics. So what can we do about this situation? We'll consider solutions at two levels: first, structural changes that could take place at multiple levels of government, and second, engagement strategies primarily appropriate at the community level (table 11). Actions at both scales are necessary.

Increase Transparency

Although it may not seem like a radical innovation, a starting point is for citizens to have the ability to tell what decisions are being made when, where, why, how, and by whom. By historical standards this is, in fact, radical.

Transparency requirements—sometimes known as "Sunshine Laws"—help make such information easily available and rule out decision-making behind closed doors except for sensitive legal matters or labor negotiations where public disclosure would undermine the government's position. Public agencies must post meeting notices, agendas, and background materials well ahead of time and make minutes and other materials available afterwards. Often meetings are televised or streamed. Routine online broadcast of public meetings increased exponentially during the Covid-19 pandemic, increasing both participation and transparency (though communities need to be mindful of the digital divide in which many low-income and rural residents lack good Internet access or computers or mobile devices). Good websites and easily searchable databases of information also help improve public sector transparency.

In California, the Brown Act enacted by the state legislature in 1953 requires all governmental commissions, boards, and councils to meet in publicly accessible facilities, hold public votes, post agendas seventy-two hours before meetings, and allow public comment on agenda items.

Strategy	Description
Structural Change	
Increase transparency.	Make decision-making processes clear and accessible; notify the public ahead of time.
Reduce conflicts of interest.	Require officials to disclose financial and personal connections to topics under consideration and to recuse themselves from votes when conflicts arise.
Restrict lobbying.	Require registration of lobbyists; restrict and/or require disclosure of contacts between officials and stakeholders with a financial interest; ban lobbying by for-profit corporations.
Improve elections.	Minimize role of private money; limit spending; enact public financing; maximize voter participation; eliminate gerrymandering of districts; reduce social media interference.
Strengthen public oversight of media.	Restrict the influence of private capital over media. Break up monopolies. Expand publicly funded media. Require fact-checking of news media.
Educate ourselves to be good citizens.	Emphasize critical thinking skills, problem-solving, and the ability to identify fake or biased news. Practice skills of understanding different mindsets within a pluralistic society.
Local Public Engagement	
Improve public hearings.	Improve the quality of hearings through clear notification, good preparation, careful listening, time limits, and facilitator identification and articulation of underlying values.
Make engagement more interactive.	Run activities in which participants interact in smaller groups, with information carefully recorded and analyzed. Includes World Café events, walking tours, design charettes, and pop-up workshops.
Appoint citizen juries.	Use groups of randomly selected citizens to help make policy.
Use old and new media more effectively.	Employ local newspapers, radio, and TV to publicize processes and issues. Cultivate relationships with reporters. Use online sites, networks, and technologies to involve the public. Explore crowdsourcing techniques using the public to gather data.
Employ participatory budgeting.	Give a modest amount of funding to local neighborhoods for residents to allocate to local priorities.
Involve youth.	Involve youth in implementing and monitoring sustainability programs as well as in representing the needs of communities.
Strengthen community institutions.	Work through community institutions such as churches, mosques, civic clubs, neighborhood associations, and festivals and other events.

Later laws prohibit a majority of public board members from gathering anywhere to discuss public business without prior notice, require local governments to make documents freely available to the public, and require public officials to disclose financial conflicts of interest. Although this can occasionally be a problem if three members of a five-person city council find themselves together at a holiday party, it has been a godsend for community groups wanting to know what local government is up to.

Reduce Conflicts of Interest

A related strategy is to reduce or eliminate conflicts of interest. This may seem common sense—why allow leaders to make decisions on items in which they have a financial stake?—but is surprisingly rare in government. Conflict-of-interest regulation requires officials to disclose tax returns, list sources of income and potential conflicts, and recuse themselves from votes related to their own past or present financial dealings. It can prohibit elected leaders from voting on any issue related to campaign contributions they've received. The Trump administration's blatant use of government for private gain showed the importance of such regulations.

Officials can also be prohibited from working for the industries they previously regulated after leaving office—that is, going through the "revolving door" between industry and government. Individuals who worked in industry leadership positions can be likewise prohibited from taking public sector positions regulating those same industries. Such commonsense changes can reduce incentives for personal gain to influence public policy.

Restrict Lobbying

More problematic is the question of lobbying, the process through which companies or interest groups seek favorable treatment from elected officials or public sector staff. Questions of free speech arise. Is it an infringement on free speech to prohibit representatives of for-profit industries from contacting officials who regulate them? Should nonprofit civil society organizations also be prohibited from making such contacts, or just the private sector? Is it enough simply to require lobbyists to report their activities, as has been done in the US since the 1970s, and can such reporting be monitored effectively? We think not.

For starters, transparency measures require that lobbyists register with the government they are attempting to influence and report their lobbying contacts quarterly so that the public can see what's going on. Many countries including the US, Canada, Germany, Poland, and Lithuania have such registration systems. However, many lobbyists don't bother registering, and few consequences exist if unreported activity is discovered. Following a 2011 scandal in which reporters for the *Sunday Times* posed as lobbyists and gave bribes to four members of the European Parliament in exchange for official favors, several countries including Austria have enacted tougher, mandatory registration in which lobbyists must report contacts with officials, issues discussed, and value of lobbying contracts. The intent is to make this information available online.[9] Whether this will make a difference remains to be seen.

A more powerful option would be to restrict the amount of money that organizations or individuals could devote to lobbying activities and campaign contributions each year. Indeed, US nonprofit corporations chartered under Section 501(c)(3) of the federal tax code are prohibited from spending more than a small amount on political activity. However, many other types of incorporated groups face no restrictions. The *Citizens United* Supreme Court decision in 2010 allows corporations virtually unlimited ability to spend money to influence government. Perhaps now is the time to say that no organization of any type can spend more than a relatively small amount in trying to win over government to its own interests. It is also time to push for publicly funded elections so that candidates can spend less time courting corporate donors and can focus on the public's needs.

Improve Elections

Free and fair elections should be a no-brainer for democratic government at any scale. Countries are slowly getting better at this. But hugely unequal amounts of campaign spending, misleading political advertisements, voter suppression, gerrymandering of districts, manipulation of social media to support particular candidates or points of view, and outright fraud are still distressingly common.

The hopeful news is that strategies abound to address such challenges. To combat low voter turnout, Australia requires citizens to vote, with a fine of almost AUS$80 for not voting. Voting takes place on a Saturday and is made into a party, with barbecues at the polling places.[10] To avoid cumbersome voter registration procedures, Sweden makes registration automatic, and voting cards are sent to all citizens three weeks before each election. If the voter has no ID, another person is allowed to verify their identity.[11] Estonia allows online voting with a required national ID card.[12] Countries such as Canada, Chile, France, Greece, Japan, South Korea, and Belgium sharply limit campaign contributions; in Canada, for example, individuals can give no more than $1,100 to a parliamentary candidate.[13] Austria, Britain, New Zealand, Slovakia, and Italy have overall limits on campaign spending.[14] Germany gives public funding to any political party that can gain at least 5 percent of the vote in any national election. Improving elections in such ways does not ensure wise choices by voters but is certainly a start toward better democracy.

The American electoral system has many equity implications, since smaller, whiter, and rural states are given outsized weight because of their importance in the Electoral College and the Senate. The ability of state legislatures to gerrymander districts to ensure that one political party dominates in each state and in that state's delegation to the House of Representatives has likewise been highly inequitable. Recent elections have made it clear that these systems need to be reexamined, for example by eliminating the Electoral College and having legislative districts drawn by bipartisan commissions.

Strengthen Public Oversight of Media

How do societies ensure that the basic information needed to make good decisions is available to the public? One obvious strategy is to have media publicly funded or controlled, which was what happened in most countries during the early days of radio and television. Commercial broadcasting was dominant from the beginning only in the United States. In contrast, since 1922 the public service British Broadcasting Corporation (BBC) has pursued its mission to "inform, educate and entertain." Most of the BBC's funding comes from the government, but its staff has editorial independence. The organization has operated as many as four television channels plus radio and Internet services, offers programs in up to twenty-eight languages, and does not allow advertising.

Other nations such as Germany, France, Sweden, Brazil, Canada, Japan, and South Korea also offer multiple public television channels in addition to commercial stations. Denmark has six national television channels and eight national radio channels. In contrast, US public TV and radio networks are weak. The Corporation for Public Broadcasting has a decentralized structure, limited public funding, and increasing reliance on corporate sponsorship. Right-wing politicians routinely seek to defund it, arguing that it reflects liberal points of view. (A counterargument is that reality has a well-known liberal bias.) National Public Radio has been a democratic voice in the US and an important source of information for millions but has faced declining federal funding and conservative criticism.[15] Strengthening these networks and launching other public media institutions would be one main way to make available good information to support US democracy.

Better regulation of for-profit media, for example by requiring broadcasters to fact-check their news and include public interest programming, would be another. Many initiatives between the 1920s and 1950s—including the Federal Communications Act of 1934, the Fairness Doctrine (1943), and numerous steps to break up media monopolies—sought to ensure that US media were not solely driven by the search for profit.[16] However, in recent decades this goal has been forgotten, and Federal Communications Commission oversight of the media has been weakened. New federal legislation on this topic that includes social media could help ensure that citizens have the information necessary to create a more sustainable society. (In 2020 social media corporations finally started to flag disinformation, including tweets by the president of the United States.)

Steps to reduce monopoly ownership of media outlets could help as well. Six conglomerates—Comcast, Walt Disney, AT&T, 21st Century Fox, CBS, and Viacom—own 90 percent of for-profit US media.[17] Even the remaining 10 percent is highly concentrated. One company within this small remaining fraction, iHeartMedia (formerly Clear Channel Communications), owns at least 850 US radio stations as well as billboard advertising in twenty-five countries. Its centrally produced programming, including talk radio shows by conservative hosts Rush Limbaugh, Glenn Beck, Meghan McCain, and Sean Hannity, thus gets an enormous audience. Breaking up such media conglomerates, as well as online media monopolies such as Facebook and Google, is an obvious place to start. The 2020 lawsuit by the FTC, backed by forty-six state attorney generals, alleging that Facebook illegally maintains a monopoly because of its ownership of Instagram and WhatsApp, is an important step.[18]

A more radical strategy would be for governments to eliminate for-profit ownership of the media altogether, on the grounds that this important educational medium should not be beholden to corporate interests. Publicly funded or nonprofit media would then be responsible for meeting the informational needs of democratic electorates. The examples of Great Britain and other countries show that conditions could be established under which publicly funded media outlets have editorial independence and sufficient funding to serve as a "democratic check."

Educate Ourselves to Be Good Citizens

Moving toward a more sustainable world requires educated citizens who can see falsehoods and hate speech for what they are, deconstruct different points of view, understand systems of influence, recognize the importance of history, challenge structural racism, and get involved constructively in civic life. Although none of us are perfect, these abilities can be practiced and nurtured starting in early childhood. To move toward this goal, a number of countries around the world have revised educational curricula to promote "critical citizenship."

Different approaches exist to this goal. In the classroom many educators seek to nourish critical thinking—an open-minded, self-directed analysis of evidence using reason in logically consistent ways. Others push for a broader "critical pedagogy" that applies critical thinking to community challenges emphasizing social justice and environmental protection.[19] The Brazilian educator Paolo Freire, who famously advocated a "pedagogy of the oppressed," was among the best-known advocates of the latter viewpoint.

Chile provides one example of a nation's evolving approach to educating citizens. Under the military dictatorship of the 1980s, "civic education" was narrowly defined as knowledge of government institutions and was taught by itself as a separate subject. After the return of democracy in 1990, the nation transitioned toward a much broader "citizenship education" in the 2000s aimed at producing free, competent, and socially responsible citizens. Students explored how human rights and democracy had evolved historically. Revised curricula in the first eight grades emphasized thinking skills, communication skills, knowledge of communities, and values of pluralism, solidarity, respect, and human rights. School governance and school climate received attention within the classroom.[20]

The goal of democratic education would be to help produce what political theorist Benjamin Barber has termed "strong democracy"—strong civil societies with not just a veneer of democratic institutions ("thin democracy"), but high levels of voter participation and citizen engagement and deliberative political discourse.[21] John Dryzek's view of "deliberative democracy" is a similar concept, requiring an educated public that can listen to different points of view and come up with collaborative and creative solutions that meet people's collective interests.[22] A related aim would be to increase "social capital"—networks of trust, reciprocity, and cooperation within society.[23] Those networks come about not by accident but through culture, practice, opportunities to engage, and the rebuilding of trust among groups. Schools, clubs, professional associations, and nonprofit organizations give citizens experience in collaborating with others and being engaged in local communities. People's skills at doing so improve, social capital improves, and democracy becomes stronger.

PUBLIC ENGAGEMENT TOOLS

In addition to such big-picture structural changes, many specific strategies can help improve public engagement at the local level. These strategies are equally important to high-functioning democratic societies. New technologies offer exciting opportunities for more inclusive and ongoing citizen engagement.

Improve Public Hearings

The venerable public hearing, in which members of the public are allowed to make comments to a legislative body such as a city council or agency board of directors, is an unavoidable aspect of democratic local government but needs to be reimagined. These hearings are often lengthy and boring affairs. Officials may not take comments seriously, leaving the room or reading while members of the public are talking, placing this form of engagement far down on Arnstein's ladder as a tokenistic method of public participation. Hearings are also frequently dominated by NIMBYs and other project opponents who are wealthier, older, and less diverse than the public at large.

Planning Commissioners Journal conducted an open survey on how to improve hearings and received 950 comments from practitioners.[24] Recommendations focused on officials being well prepared, stakeholders being effectively notified about meetings, and the process being clearly explained and respectful. Respondents emphasized that officials should actually listen to members of the public, rather than reading papers or talking to aides, and should keep an open mind on the issue at hand. The Institute for Local Government suggests that the chair of meetings play an active role by focusing on and articulating the values underneath public comments and by making the trade-offs of different policy directions clear to all. To keep the process moving, time limits on comments are usually necessary, and officials can ask respondents simply to say, "I agree with————" rather than repeating arguments. They can also ask if anyone in the room represents viewpoints that have not been heard and can point out constituencies that are not present, such as members of minority groups, people in other jurisdictions, future generations, and natural ecosystems. Live streaming of public hearings means that many of those interested do not need to attend in person but can still learn from the events by watching at the time or later. Through technology cities can come up with ways to get feedback from those participants as well.

Active facilitation of public meetings by a government official or professional facilitator is important in order to help participants learn from the process and feel that their time has been usefully spent. The facilitator can demystify the process for participants, make sure that everyone is heard, make sure that nobody dominates the discussion, record and emphasize points of agreement, and try to keep the process lively, interesting, constructive, and fun. These are significant challenges, but ones that a patient, skillful professional can address. Training can help, as well as humor and a sense of perspective. There is nothing better than a good laugh to defuse tension in a room and help people put the situation in perspective.

Make Engagement More Interactive

Even the best-run public hearing is a passive affair, with members of the public sitting in chairs listening to speakers. More active modes of engagement are increasingly common.[25] In focus groups—facilitated discussions of a topic involving perhaps a dozen community members—participants are able to engage one another in exploring the intricacies of an issue. Rather than just stating a position, people work together to develop creative solutions. By observing or participating in the discussion, public officials can understand perspectives on the issue at greater depth than otherwise.

Stakeholder dialogues are a similar process in which city staff or consultants set up meetings between public officials and different stakeholder groups, for example environmentalists, local business owners, civic organizations, historic preservationists, neighbors, and low-income individuals. The aim is to really understand stakeholders' perspectives, and also to solicit their suggestions and buy-in. Groups are typically kept small, by invitation only, so as to achieve intensive, high-quality dialogue about the topic.

Community conversations are larger, open events with breakout groups in which participants discuss an issue and record their concerns. The large meeting then comes back together, with the breakout groups reporting back their questions and points of agreement, and facilitators carefully recording and later analyzing the entire range of perspectives. The World Café method, within which participants move through multiple rounds of small-table discussion focused on particular questions, is one version of this technique.

Other active engagement mechanisms include scenario planning exercises that help participants understand the implications of policy alternatives, pop-up workshops in which members of a planning staff set up displays in locations within the community and solicit input, and community walking tours during which participants contribute their personal knowledge about a site or neighborhood to a leader or each other. Walking tours are particularly useful in developing understandings of how community spaces actually work for people. By being in those spaces, participants may be able to develop constructive suggestions for improvement that they wouldn't think of while sitting in chairs in a meeting hall. Virtual Reality (VR) may also provide opportunities to show people existing and proposed conditions. Online community dialogues can provide a forum for community discussion. Such online communications may need to be carefully moderated in order to keep discussions constructive and avoid them being dominated by opinionated individuals.

Design charrettes are another long-established urban design mechanism for active public engagement. Typically a team of outside consultants leads an intensive weeklong process with community members to develop a vision for a place. After an initial brainstorming session the professional team takes the public input and develops some initial design alternatives. They present these back for feedback and then produce a second cut, arriving at final proposals by week's end. The success of this method depends on the skill of the consultants in incorporating public input and developing proposals that represent a true integration of community ideas rather than just a repackaging of predetermined concepts. Charrettes are time-consuming both for the professionals and for community members. In recent decades, they have been used so extensively within some communities that they lead to "charrette fatigue." Still,

FIGURE 12. Many types of interactive public involvement exercises have been used in recent years to brainstorm improvements to cities and neighborhoods. (S. Wheeler)

they can be a very productive form of engagement, and the often-lovely perspective drawings and site plans they produce can be a useful way to convey place-making ideas to the broader public (figure 12).

Appoint Citizen Juries

Historically, local governments have often set up advisory boards as a way to get public input on issues. These were often starchy Blue Ribbon Advisory Councils composed of heads of selected stakeholder organizations, and at times they have been accused of being rubber stamps for official policy. However, carefully chosen citizen steering committees can still be useful to secure a wider range of community input to proposed actions. In Philadelphia, the Citizens Planning Institute is the outreach arm of the city planning department; it educates and engages community activists, building local knowledge and capacity to more effectively engage with official processes.[26]

A different approach is represented by the "citizen jury." In this model, a city randomly selects a group of residents to advise it on important matters. These need not be people with particular expertise, residents with long-term experience in the city, or even citizens of the country involved. Rather, they are simply folks who happen to be living in the community at the time. The jury is convened, briefed, given full access to city materials, and instructed to produce recommendations on certain topics. To work well, the process must be carefully coordinated and be able to directly affect policy.

The city of Melbourne, Australia, for example, randomly selected forty-three residents in 2014 to form a People's Panel to advise the city council on a $5 billion, ten-year financial plan. To reflect the importance of business to the city, half of the panelists were business owners and half were other urban residents. Coordinated by an NGO called the New Democracy Foundation, the group met over four months with complete access to the city's financial information and assistance from any outside experts it chose. The group achieved more consensus than anyone expected and made eleven recommendations to the city council, which incorporated many into a plan finalized the next year.[27] Such radical reenvisionings of local democracy can avoid the problems of entrenched stakeholder interests and government institutions found in societies worldwide.

A related citizen jury approach has been used in some states and countries to redraw the boundaries of congressional districts. Historically US Republican and Democratic parties have often used their control over state legislatures to gerrymander district boundaries, maximizing the number of their own candidates elected to Congress or state legislatures. The results subvert democracy. For example, during the 2018 election North Carolina citizens split their votes almost evenly between Democrats and Republicans, but because of gerrymandered districts Republicans won three-quarters of congressional seats. In response, states such as Iowa and California—as well as countries such as the United Kingdom and Australia—have appointed nonpartisan citizen commissions to draw new legislative boundaries. California adopted such a procedure through a ballot referendum in 2008. Nearly thirty thousand citizens applied in 2009 to serve on the commission. The state auditor's office oversaw a process combining random selection, interviews, review by leaders of the major political parties, and further random selection to arrive at a panel of fourteen highly capable citizens without a majority for either political party. The new district boundary map this board created led to legislative and congressional seats being held in more or less equal proportions to the popular vote. Score one for democracy.

Use Old and New Media More Effectively

An essential local leadership skill is to use various types of media to engage the public. There are many ways to do this, and technologies are constantly evolving that can improve civic media. Planners and community groups can develop relationships with reporters for traditional news media and feed them stories. They can meet with editorial boards and stage events to dramatize urban planning decisions. They can run ads about public issues. They can ask citizens to upload digital stories about local conditions. Potentially cities can support independent radio and TV stations, for example by giving them public facilities. Where local media do consistently cover local and regional planning issues, as in Portland, Oregon, this has helped contribute to an effective urban planning culture.

Local governments can also make their websites more easy to use, engaging, and informational for citizens. Separate, easy-to-find portals such as the "311" sites set up by many large US cities, or OpenDataPhilly, a metaportal for that city, can help residents locate online sources of information. Comprehensive and easy-to-use GIS portals for citizens to see and download spatial data are helpful as well. Videos, games such as Sim City, visualizations of future development, and online planning-related activities for children could help engage constituencies who would never otherwise think about planning subjects. Governments might need to hire new staff to run these electronic communications. They could also partner with schools to help create content. But the results would be worth it in terms of education, citizen engagement, and inclusive planning.

Many efforts have been made worldwide to set up neighborhood-based electronic portals such as NextDoor that can be used both by residents to network with one another and by public officials to engage people in decision-making. However, like in-person neighborhood associations these tend to be dominated by longtime residents who own property, rather than younger, less well-off, or less educated individuals. Members of minority groups or recent immigrants may not feel welcome or comfortable using such portals. The information shared on these websites also tends heavily toward mundane needs such as finding lost cats or good home repair personnel, rather than meaningful, in-depth discussion of public issues. Academic research suggests that social media are best thought of as an adjunct to traditional public participation methods, and that they may be more useful in some ways rather than others, for example in collecting grassroots spatial data on a community and in promoting local self-organization.[28]

Still, municipal governments can gather large amounts of data from citizens voluntarily over electronic media, allowing people's daily experience to help run the city. For example, Boston has tested an app named Street Bump that lets drivers' phones automatically report potholes to authorities whenever their cars hit them, using the phone's accelerometer and GPS. Moscow has developed an app called Active Citizen to gather public opinion on issues, conducting 2,800 app-based polls with two million citizens participating. Topics include changes to bus routes, expansion of pedestrian zones, and the naming of metro stops.[29] On a more ambitious scale, Helsinki employed Public Participation GIS software to get resident input on potential urban infill or urban greening locations during its Master Plan process. A total of 3,745 residents marked 32,989 potential locations for such initiatives on a digital map of the city. This number of respondents was far more than would have shown up in person to public meetings, and the data helped give planners a good sense of public sentiment on these topics.[30]

Employ Participatory Budgeting

In 1989 the city of Porto Alegre in Brazil pioneered a radical strategy known as participatory budgeting, through which local neighborhoods were given the ability to decide how to spend some 15 percent of the city budget. In a yearlong process citizens negotiated priorities among themselves and with government staff. More than 120 of Brazil's 250 largest cities subsequently adopted this approach, often coupled with quality-of-life indicators that allocated additional resources to poor neighborhoods. An academic analysis has shown that Brazilian participatory budgeting is associated with increases in spending on health care and civil society organizations and decreases in infant mortality rates.[31]

The city of La Plata, Argentina, has also engaged in participatory budgeting since 2008, using a combination of online and offline methods to allow citizens to have input into budget allocations. Annually, more than 10 percent of the city's population participates in this process, affecting more than 6 percent of the city's budget.[32] Ukraine has also used this tool, at times in combination with general assemblies of local residents and public opinion polls.[33] Chicago, Boston, and Oakland have experimented with small participatory budgeting programs. New York City initiated participatory budgeting in the mid-2010s, giving up to \$31 million a year to local council districts to allocate as citizens see fit. These funds catalyzed small-scale local projects such as street repairs and solar panels and rainwater catchment systems on school roofs.[34] An NGO named the Participatory Budgeting Project is promoting the idea worldwide.

Involve Youth

Getting young people involved in local government can be extraordinarily educational for future leaders as well as providing useful input and person-power to local governments. A number of K-12 schools worldwide make civic activism part of their curriculum, and some nations have taken the lead on this as well. Peru is one of the most active. The Programa de Embajadores Perú Agenda 2030 program, initiated in 2016, has worked with more than twenty different organizations to involve youth in promoting national action around the UN's Sustainable Development Goals. In the program's first years more than 162 young people helped conduct fifty-seven community workshops and represented local community needs in meetings with UN staff.[35]

Strengthen Community Institutions

Engaging the public through traditional institutions such as churches, synagogues, mosques, civic clubs, neighborhood associations, environmental groups, schools, festivals, farmers' markets, sporting events, and the like is a tried-and-true method of outreach. Many residents who might never know otherwise about urban planning issues could learn about them in this way. The downside is that such outreach can take an enormous amount of time. Staffing tables or sending speakers to numerous local events is more than many cities are capable of. However, if local leadership is enlisted to help disperse information and solicit feedback through events or electronic networks, outreach can potentially be maximized with a modest public sector outlay.

A strong ethic of social inclusion and concern for the least well-off can help a municipal government gain respect and engagement from its residents. A leading global example is Vienna, Austria. That city's routine placement at the top of global quality-of-life rankings reflects the government's commitment to affordable housing, public green spaces, inexpensive public transit, and extensive citizen engagement.[36] Municipal authorities own a quarter of the housing stock, build large numbers of affordable units using national and local funds, and provide housing subsidies for half of all residents. The city keeps transit costs to the equivalent of $1 a day for annual passes. In 2011, Vienna launched a "Smart City Initiative" that brought large numbers of stakeholders together to plan sustainability-oriented actions around the themes of human development, environment, administration, economy, energy, and mobility. It differed from other "smart city" initiatives worldwide in that organizers made clear that a "focus on people" would seek to prevent an emphasis only on technical solutions. The city council adopted the resulting Framework Plan in 2014. Progress toward the plan's fifty-one objectives is monitored by a team of 120 people from city government and external stakeholders. The city has already achieved certain goals such as reducing GHG emissions by 35 percent compared with 1990 levels.[37]

CONCLUSION

The twenty-first century is a challenging time for democratic decision-making and public engagement at all levels of governance. Private power and wealth have undermined democracy in many countries. The public is besieged with massive amounts of information, much of it misleading or false. Traditional news media have declined.

Rising to the challenge will require both structural changes in societies and local day-to-day strategies. More transparent decision-making, conflict-of-interest regulations, election and lobbying reform, better public oversight of the media, and better processes to educate citizens are among the large-scale structural changes necessary. At the local level, improved public process, better use of social media and new technologies, interactive events such as charettes, and use of innovative tools such as citizen juries, citizen planning institutes, and participatory budgeting can help. Thinking outside the box on both these levels can strengthen social ecologies and help more functional public decision-making to occur. Making democracy work in these ways will be essential in order to move toward more sustainable and inclusive cities.

How Can Each of Us Help Lead the Move toward Sustainable Communities?

Each morning many of us go to work or school wanting to change the world. We are painfully aware of contemporary problems such as global warming, social inequality, racial disparities, gender discrimination, environmental injustice, overconsumption, underdevelopment, and waste of natural resources. Over our newspaper or social media feed we shake our heads at the insanity of the current world.

Sometimes we can't bear to read or hear any more. We want to do our part for social change. But it is hard to know how to do that given the scale and magnitude of the problems we face.

Many of us who have dedicated our lives to making a difference find ourselves sitting at desks wishing we could do more. Maybe we are working on reports that we worry no one will read. Or we are asked to approve a project that we don't think has a strong enough focus on social or racial justice. Or we work for a small NGO and worry that no one will listen to our group's perspective. We try our best to work hard and cultivate an ethical approach to our professional work but are concerned that our efforts aren't doing enough, or even, as with green gentrification, that we are making things worse in some ways. We are stuck operating in a system that we know is designed to reproduce inequity. So how do we ethically do our work while promoting change?

It can be depressing to be in such positions. In many cases bosses or clients appear wedded to BAU. There appears to be no money or political support for radical change. Social, racial, and environmental concerns often take a back seat to economic growth. Literally or metaphorically, we feel trapped in a cubicle, unable to make the systemic change we know is necessary in a world that is rapidly warming and that is overwhelmingly unfair.

How do we move beyond such feelings of powerlessness?

There are no easy answers to this question. But there *are* answers. Though the process can be time-consuming and may depend on slowly gaining skills, experience, and greater authority within our organizations, each of us can find niches that help us do our part in creating more sustainable and equitable communities and a more sustainable world. The current global challenges need strategic, creative optimists who are willing to demand more radical approaches and identify pathways for action.

In this final chapter, we'd like to focus on two things: ways that professionals in public, nonprofit, and private sectors as well as citizen activists can help bring about more sustainable, inclusive, and just communities on a day-to-day basis, and ways to help shape social ecologies in more constructive directions in the long run, so that our work and that of others can eventually bear fruit. We build off successive efforts and social movements. As Martin Luther King Jr. reminded us, "The arc of the moral universe is long, but it bends towards justice." What role will we each play in our personal and professional lives in this ongoing struggle?

PAST TYPES OF ACTIVISM

Generations of creative individuals have struggled with the question of how to bring about change. These activists might be grouped into several categories based on their primary role: visionaries, reformers, organizers, and implementers.

Visionaries come up with radically new ideas or combinations of ideas and get others excited about them. They see pathways in which their ideas can make changes to the existing paradigm. At their best, they become influential and well known. Often, however, they are ignored or dismissed as idealists or unrealistic.

Reformers pursue more incremental change to existing systems and typically work within existing institutions. They can potentially make a significant difference by improving already-established systems. The risk, though, is that these changes will be small, that they will not challenge systematic injustice, and that the bigger-picture vision of change will get lost.

Organizers help coordinate activities and build momentum, either inside or outside institutions. Traditionally grassroots community organizers work with residents of a given neighborhood or community to identify community needs, build political strength, and get those needs met. But there are many other types of organizers and policy entrepreneurs who work within and across governments, corporations, NGOs, or the media to get things done. Organizers are typically "people people" who enjoy interacting with others and networking. They are essential to build momentum and political power. However, if their efforts are not well connected to visions of change, strategic thinking, and well-thought-out proposals, they may not actually help communities move toward sustainability. Or they may be effective at organizing at a local level but less effective at influencing more systematic change.

Finally, implementers are good at putting changes into practice, whether through devising and running programs, developing policy, designing and constructing places, or other means. They are typically pragmatic individuals who like making systems run well. Very little lasting change can happen without them. However, without a good connection to these other modes of change, they can produce well-functioning societies headed away from sustainability rather than toward it, or can implement policies that do not get to the root of problems.

Most of us combine more than one of these roles or engage in different roles at different times. It is important to be aware of one's own character, talents, and weaknesses, pick opportunities that work well with that mix, and improve one's effectiveness over time. Specific ways of working for change are continually appearing as cultures and technologies change, more people realize the limitations of BAU, and crises such as global warming become more severe. In addition, as planning and design professions open up to people from different backgrounds, these new participants bring with them new perspectives, problem framing, cultural competencies, and ways of working. Having more women involved in planning has introduced issues and values that were not previously emphasized, especially those related to people's daily lives. As more people of color enter leadership roles in government, civil society, and business, they also bring with them different perspectives on institutions. For instance, rather than seeing the police as protective agents, many people of color can recount years of systemic racism at the hands of the police. They may demand that we reevaluate the role of law enforcement and allocate resources in ways that better support nonwhite communities.

The vision dimension of social change, the focus of this book, is particularly important in the twenty-first century in order to develop alternatives to the neoliberal institutions, structural racism, modernist worldview, and global capitalism that have dominated the world under BAU. Whatever our personal character, inclinations, heritage, and background, attention to big-picture, structural analysis and vision is crucial now. We can all agree that BAU is unjust and not sustainable. Throughout history many of the most creative visions about what we would now call sustainable communities have come from those outside mainstream behaviors and institutions. Examples are numerous, but let's look at a few.

Ildefons Cerdà, mentioned in chapter 8, was a middle-aged Spanish civil service engineer in the early 1850s when he chose to leave his secure government job and develop his ideas for greener, more diverse new urban communities. It helped to have inherited some family money, but what was important was his decision to give himself some space for reflection and his patient development of a well-thought-out vision for improving urban quality of life. He put his proposals forth through books, exhibits, graphic designs, and eventually legislation after winning election to the Spanish parliament. The centerpiece of his work—the Eixample district of Barcelona—was actually built and remains one of that city's most attractive neighborhoods.

A generation later, Jane Addams was a depressed twenty-seven-year-old unable to pursue a medical career because of ill health when she and her partner Ellen Gates Starr decided to open the first American settlement house. They had seen British models of these centers where people of all social classes could mingle and immigrants could gain knowledge and skills while also celebrating their own cultures. Soliciting donations from affluent Chicago residents, Addams opened Hull House on the city's inner west side. This facility eventually expanded to thirteen buildings and became a model for hundreds of other settlement houses in North America over its seventy-year lifetime. Stymied in her initially chosen direction, Addams had managed to create a meaningful role for herself by adapting creative ideas from elsewhere to a new context.

A couple of generations later, a young, degree-less writer working at *Architectural Forum* magazine in the 1950s decided to write essays about the design of cities. This decision was influenced by her role as a local neighborhood leader fighting a freeway that New York City's planning czar Robert Moses wanted to bulldoze through Greenwich Village, where she and her family lived. This enterprising activist, whom

FIGURE 13. Greta Thunberg urges the European Parliament in 2020 to take action on the climate crisis. (European Parliament/Creative Commons.)

we've encountered before, Jane Jacobs, published a series of magazine articles culminating in her 1961 book *The Death and Life of Great American Cities*. Jacobs had figured out a way to use her writing skills and down-to-earth common sense to communicate radical ideas to multiple generations of readers. In doing so, she fundamentally challenged the planning profession to think about the public sphere differently.

Closer to our own time, Annie Leonard worked for twenty years as an organizer with nonprofits such as Greenpeace before coming up with a breakthrough concept: a short video about the system that produces runaway consumerism. With Leonard talking straight to the camera in front of simple, hand-drawn cartoon graphics, the video explains the sustainability problems of complex economic systems in readily understandable ways. *The Story of Stuff* video went viral in 2007 and has been viewed tens of millions of times in fifteen languages, helping people worldwide rethink consumerism. It has been a particularly effective educational tool, helping K-12 children learn a new perspective on consumer society.

The recent groundswell of citizen activism against BAU provides still more examples of how individuals can develop a sustainability vision within varying professional roles. In Tanzania, Edward Loure has built a grassroots movement to establish community control over land rather than individual landownership. In southern Chile, Mapuche Indian Alberto Curamil has organized resistance to destructive dam-building projects. In Sweden, high school student Greta Thunberg launched the international "school strike for climate action" movement (figure 13). In the US, former restaurant server Alexandria Ocasio-Cortez won election to the US Congress and has been a progressive powerhouse pushing for racial and economic and climate justice. Countless less well known others have likewise made a difference both inside and outside traditional employment by developing their own visions of change and applying their other personal skills in the process.

APPROACHES TO SOCIAL CHANGE

The twentieth century was a highly mixed time in terms of promoting sustainable and just communities. On the plus side, important social movements advanced the rights of women, people of color, LGBTQ individuals, and the environment. Other movements opposed violence of many sorts and sought structural change in economic and political systems. However, the constellation of beliefs, values, and attitudes known as modernity held sway for much of the century. Societies had faith that supposedly objective experts could guide public policy and often believed that science and technology would supply the solutions to most problems. Capitalist economics spread virtually unchecked worldwide as the century neared its end. Yet a number of alternative theoretical movements helped lay the groundwork for different approaches in the remainder of the twenty-first century.

Perhaps foremost among these was continued development of Marxist critiques of political economy. For 150 years this important theoretical tradition has analyzed how elites wield power gained from control of capital in order to structure societies in their own favor. David Harvey, Manuel Castells, Doreen Massey, and Mike Davis have been among the foremost thinkers in this regard during the past half century.[1] The implication for social change activists is that understanding dynamics of economic and political power in society is an essential foundation for change, and conversely that ignoring these may lead to solutions that fail in the long run.

Ironically, officially Marxist countries succumbed to many of the same problems of capitalist BAU, in large part because of their embrace of technocratic modernity. The former Soviet Union and its satellites did not do well at either meeting human needs or protecting the environment. It is clear that there is no silver bullet approach to governing that solves all our problems. The various -isms (Marxism, capitalism, socialism, etc.) on their own will not solve the world's problems. Instead, it will take citizens and activists constantly engaging with political, social, and economic systems to "bend them toward justice."

A separate body of theory has focused on the role of institutions within societies. Various versions all emphasize the role of organizational structure, social norms, laws, rules, and political processes.[2] The implication is that for social change to happen, the institutions of a society must evolve. Change then becomes in large part a question of modifying or replacing these institutions. Institutionalist perspectives are particularly important for those of us wishing to promote long-term changes in social ecology in incremental ways. The prerequisite is getting people to admit that a serious problem exists requiring institutional change. However, with growing recognition of systemic racism and injustice, as well as other systemic problems, many people are rightly demanding large-scale institutional reforms.

A third, more personal approach is Paul Davidoff's notion of advocacy planning. This was originally advanced in a 1965 article in the *Journal of the American Planning Association,* in which Davidoff argued that "the right course of action is always a matter of choice, never of fact. Planners should engage in the political process as advocates of the interests of government and other groups. . . . Plural plans rather than a single agency plan should be presented to the public."[3] This approach laid aside the modernist ideal of social scientists being able to come up with an objectively right answer to a policy question and instead called for professionals to develop multiple alternatives for the public and then to advocate for those that they thought could best meet public needs. As advocates they would need to coordinate with constituencies inside and outside government to ensure that needed proposals were fully developed and supported. Advocacy planning requires good research and communications skills to effectively argue why particular initiatives are desirable, but also the humility to know that this is one choice among many and that better solutions may be possible.

Notions of radical, insurgent, and tactical urbanism are closely related.[4] In these, the individual works either within or outside government, often at grassroots levels, to build alternatives to an oppressive status quo. Activists may take guerrilla action to establish examples of different forms of urbanism, may create pop-up spaces to pilot new concepts or uses of space, or may work more strategically within government to structure an environment for radical change. For example, Victoria Beard shows how Indonesian public sector and private workers used grassroots community development efforts, including setting up clinics and libraries, to help develop social infrastructure that could then assist with larger-scale political change when the regime of dictatorial president Suharto was finally toppled.[5] These individuals recognized the oppressive institutional structure they were working within and carefully laid the groundwork for constructive change when the opportunity presented itself.

A further theoretical viewpoint emphasizes ways that people and societies learn. Planning scholar John Friedmann termed this approach social learning.[6] The implication is that professionals should seek every opportunity to help decision-makers, colleagues, members of the public, and themselves learn about alternative ways of addressing problems and processes of social change themselves. They might organize workshops, hearings, conferences, symposia, design charrettes, research collaboratives, site visits, walking tours, and web-based events or networks in order to promote mutual learning. This branch of theory overlaps with the communicative school of planning philosophy, which emphasizes processes of dialogue, negotiation, and consensus building.[7]

Brazilian educator Paolo Freire emphasized social learning through his philosophy of the "pedagogy of the oppressed." In this approach the professional works with groups of ordinary people to identify, understand, and become actively involved in meaningful local issues. The end result is well-rounded development of individual knowledge and conscience, a process Freire called "conscientization."[8] Freire argued that it was important for people to see that the slum communities that they lived in were not "normal" but the product of systematic oppression and disinvestment. With this understanding, they could begin to demand that the systems that oppressed them change.

When formal processes of social change prove futile, a final philosophical tradition useful to social change activists has been that of nonviolent civil disobedience. Many social movements over the last hundred years have been based on this, including Gandhi's campaign for Indian independence, the US civil rights and anti–nuclear power movements, the nuclear arms control movement of the late twentieth century, and the recent Black Lives Matter movement. This philosophy emphasizes deep commitment to values beyond those of existing institutions and the use of sit-ins, protests, and nonviolent civil disobedience to leverage political action and media attention.

Many more bodies of theory are important as well. Feminism, critiques of white supremacy, and queer theory are examples. The first of these argues for incorporating theories of gender and patriarchy into our analyses of society and challenging masculinist institutions and behaviors. It also suggests incorporating an ethic of care into urban planning and design, for example through policies to ensure decent housing, health care, childcare, eldercare, and universal design.[9] The second stresses the urgent need to consider the pervasive influence of racism within societies and to develop strategies to overcome it. The third emphasizes ways that normative concepts of sexual orientation have straitjacketed individuals and cultures and the need to move beyond those in our thinking and action.

As noted throughout this book, multiple strands of theory are useful in different ways and can be woven together to inform future-oriented action. No perspective by itself holds all the answers. But together they can help us understand the social ecologies we work within, and strategies to be effective change agents.

MINDSETS FOR CHANGE

The mental strategies necessary to help bring about more sustainable development in the twenty-first century can draw from many of the roles and theoretical traditions highlighted above. But there will be three main differences with BAU: (1) appreciating the urgency of action and the need for long-term change, (2) developing an holistic or ecological worldview that understands the world in terms of complex, evolving, intersectional systems, and 3) focusing on results-oriented problem-solving.

These flow naturally from the concept of sustainability. Things are not okay. Current situations, particularly global warming, threaten humanity's long-term survival, plus cause human hardship and ecological damage day-to-day. It's not enough to spend our careers pursuing intriguing ideas and fun technologies or else making money for ourselves or others. We don't have the luxury of doing that anymore because our collective future is so uncertain. All of us have a responsibility to use our precious time and talents to help solve problems and make the world more sustainable and just. Hopefully doing so can be reasonably fun and rewarding also, not to mention deeply meaningful. But the focus needs to be on collective action and problem-solving.

The sustainability mindset is challenging. Instead of buying exclusively into one particular mental perspective—a political ideology, a community identity, or a discipline such as economics or sociology or engineering—it is essential to understand the world holistically. Many theories and points of view can be useful. Any one by itself can be limiting. Building an increasingly holistic mental framework can help us understand the social ecology around us and develop both short-term and long-term strategies for improvement. But doing this is not easy and takes time and practice.

That being said, some ways of understanding the world can be more useful than others in any given context. Race, for example, is important for analysis of the American context given the United States' history of slavery, systematic racism, white supremacy, and white fear.[10] The election of Donald Trump, the Black Lives Matter movement, and the impacts of the Covid-19 pandemic cannot be understood without this historical perspective on race in America. A feminist perspective is also critical, since female viewpoints have often been missing throughout history. Women's stories are often untold. Their perspectives and needs have traditionally been marginalized. The same goes for the narratives and viewpoints and histories of indigenous peoples. Finally, an understanding of the history of colonization around the globe helps us to interpret and contextualize unjust conditions on the ground within many developing-world countries and to understand First World responsibility for those.

Leaving out some perspectives on any given context can lead to major sustainability problems. For example, planners have frequently relied on economic analyses to propose new business parks at the edge of cities as a way to address the need for employment or tax base. However, they failed to consider the traffic and GHG emissions these facilities would generate (mobility and environmental dimensions of the situation), the perceptual and phenomenological impacts of this motor vehicle–

oriented landscape (aesthetic and existential dimensions), and ways that such suburbanization perpetuated racial segregation and urban disinvestment (social equity dimensions). As a result, they helped create sustainability and justice problems ranging from the climate crisis and disinvestment in primarily Black urban communities to the soullessness, injustice, and lack of community within much of suburban America.

None of us can be experts in everything. But it is possible to scan any given context and ask ourselves questions such as the following:

- What stakeholders are involved at various geographic scales, and what are their needs and interests? Who is not at the table?
- What are the environmental, economic, and social dimensions of this situation at different scales? How do they interact?
- What is the history of this place or situation? How have different factors coevolved over time? How do these affect the current context and future possibilities?
- How do gender, race, class, age, sexual orientation, and ability come into this picture? How can solutions promote equity for disadvantaged groups?
- How might different theoretical perspectives apply to this situation?
- Who benefits from a particular approach? Who loses? Who decides?
- What are the short- and long-term implications of various alternatives?
- Is it possible to coordinate this effort with others to amplify our impact?
- Is it possible to think outside the box to come up with new solutions?
- Are we using the right problem frame to assess the situation?
- How can we use this situation to create a model of problem-solving that meets environmental, economic, and social needs far into the future?
- How can we measure success?

Understanding which factors matter the most in any given situation takes experience and will depend on the assumptions one makes. It's important to be up-front about these assumptions.

Doing so is the opposite of, say, travel demand models historically used in transportation planning. In that "black box" form of decision-making, experts put numbers representing existing traffic patterns plus assumptions about future population, jobs, housing, and economics into a computer, and the results helped determine which transportation systems would be built. The model itself was impenetrable to the average citizen. It was run by staff within a large agency, and relied on numerous assumptions—for example, that new highways would not in turn generate more sprawl development that would lead to more traffic—that were never fully taken into account. Externalities such as GHGs, residential segregation, or urban disinvestment were also deemed politically off the table or not relevant to the model. They were the responsibility of another department, or unrelated. The result was a logical, rational process of decision-making that nevertheless produced disastrous results in terms of sustainability and equity. To counter such problems, a strong emphasis on transparency of values, assumptions, and worldviews is needed in order to move toward sustainable cities. Intersectional thinking is a prerequisite. The officials building the models also need to be diverse enough and creative enough so that they can ask radical questions such as "Is this the right way to approach this problem?" and "What's missing in our thinking and our model?"

ON-THE-GROUND STRATEGIES

We can put these mental frameworks into practice through a number of practices (table 12). What follows is a starting point to guide future urban sustainability leaders.

Frame Initiatives as Opportunities and Link Them to the Vision

How issues are framed matters greatly. Presenting sustainability initiatives proactively as opportunities rather than as responsibilities, necessities, or chores is one useful strategy. It is more uplifting and motivating to think about making good things happen than to worry about heading off disaster. There is also power in "small wins" that can lead to larger systematic change and a sense of collective efficacy. Fear of disaster can be motivating as well but must be used with care.

Reimagining situations and getting positive visions out there—backed up by well-thought-out specifics—helps inspire ourselves and others. Often organizations, public sector processes, and social movements get

TABLE 12   STRATEGIES FOR TAKING SUSTAINABILITY LEADERSHIP

Strategy	Description
Day-to-Day Actions	
Frame initiatives as opportunities and link them to a vision.	Point out directions toward positive futures, develop inspiring visions, and create user-friendly news hooks.
Understand contexts.	Know all dimensions of the context you're working in.
Develop multiple leadership skills.	Use multiple skills such as facilitating, envisioning, organizing, coordinating, keeping track, etc.
Run effective meetings.	Be organized. Be inclusive. Set good ground rules. Keep the meeting moving. Make sure everyone is heard, including people who can't make it to the meeting. Record and celebrate points of agreement.
Develop sustainability education.	Provide opportunities for individual and collective learning. Emphasize what has been learned.
Improve implementation.	Follow through and make sure stuff gets done.
Take care of ourselves.	Anticipate difficult situations and develop strategies. Know your own skills and limitations. Keep a sense of balance.
Long-Run Ways to Help Social Ecologies Evolve	
Build awareness about the structure of societies.	Help community members understand economic, institutional, social, environmental, and technological forces leading to social evolution.
Develop vision.	Create strong, easily understood alternative futures for the public.
Reform institutions.	Articulate steps to create organizations that are more humane, flexible, and efficient.
Build networks and coalitions.	Bring organizations and people together to collaborate on ways to move from powerlessness to power.
Expand civil society.	Build nonprofit organizations that can provide essential functions and advocate for public and private sector change.
Build diversity.	Proactively include others from different backgrounds.
Nurture creativity.	Support production of images, graphics, phrases, and ideas that show how the world could be different.

caught up in day-to-day details. The big-picture vision gets forgotten. Bringing it back in can energize people and get them away from arguing over details. Also, many individuals either have never thought about big-picture change or think that such visions are impossible. Holding up a long-term vision and then spelling out steps to get there can help people see that fundamental change is actually a feasible goal.

Framing opportunities is often best done by presenting a range of alternatives or scenarios, from the moderate to the radical. This tactic allows people to get used to more radical ideas slowly and allows everyone to compare multiple approaches and perhaps even invent a new synthesis that is better than any previously articulated option. Through deliberation, the goal is for stakeholders to imagine new possibilities, new combinations, and new paths forward that go beyond what they could think of individually.

It is important to get a truly radical alternative into the initial mix—one that will in fact accomplish sustainability goals such as achieving carbon neutrality or dramatically improving social and racial equity. But people may then need time to get used to this wild-sounding idea and see it as possible. The community may also need to take incremental steps toward it for practical or political reasons.

Showing how multiple goals can be met by a certain action is also useful as a framing strategy. Because of their holistic nature, sustainability initiatives usually do try to satisfy multiple needs, whereas BAU approaches focus more narrowly on a particular perceived requirement. An urban growth boundary, for example, can be presented, not just as a way to save farmland and open space, but as a strategy to help revitalize downtown, since it will encourage new homes and businesses to be located there rather than on the urban fringe. It can also help improve recreational opportunities and ecological health. A tax reform proposal can be presented, not just as a way to raise revenue for much-needed parks, schools, and community services, but as a way to save future costs and promote social equity (if structured so that wealthy taxpayers pay a higher rate than less affluent ones). A climate change plan can be presented as a way to build a green economy, promote local green jobs, reduce energy costs, and improve energy security as well as to help meet global GHG reduction goals.

Framing that links into emotional and newsworthy hooks is often essential in order to get sustainability concepts into the mainstream media and public consciousness. This may mean using particular keywords or phrases that the public can relate to. The early-2010s Occupy Movement's "1 Percent" framing is a good example; the vast majority of the public can relate to being part of the less affluent 99 Percent, and the idea that it is wrong that 1 percent of the population has hijacked most of the benefits of economic growth is intuitively understandable. Effective framing may also mean linking into timely public debates or policy handles. Launching a new initiative immediately before a legislative session can help build momentum for policy change. Piggybacking on a major event such as a hurricane (for climate change initiatives) or a prison riot (for penal justice reform proposals) can leverage attention from the media and policy makers. Political scientist John Kingdon's work is helpful for those interested in capitalizing on what he terms "policy windows" that open after crises.[11]

Finally, individual and institutional self-interest is often a useful framing strategy with particular stakeholders. Politicians and heads of organizations usually like to think of themselves and their communities as leaders. Showing them how they can make a name for themselves (and get reelected) by embracing cutting-edge ideas that promote sustainability and equity can be a way to get them to take leadership. Showing how a city or town can become known as a sustainability leader (and get positive press, tourism, economic development, or business from that) can be a way to encourage both leaders and residents to buy into such an initiative.

Understand Contexts

The starting point for developing sustainability strategies is usually to understand the context of any problem as deeply, holistically, and intersectionally as possible. One must spend time researching the issue or place in question and understand the historic context. If the challenge is place-based, it is important to spend time traveling through the locale, observing it, identifying constituencies, talking to people, looking at data, and piecing together the picture of how it has evolved in the past and might change in better ways in the future. It is particularly important to listen carefully to people's stories. People need to be heard. Planners need to hear what people are "really" saying, understand what they need, and empathize with the dreams they have for themselves and their communities. Then they need to come up with alternative futures that allow these dreams to come to fruition.

Twentieth-century officials often didn't do this. Drawing on abstract studies, unfounded assumptions, and design or planning ideology, public sector leaders often rammed through urban renewal and modernist redevelopment plans that targeted and destroyed historic neighborhoods and displaced vulnerable communities, particularly those with low-income individuals and people of color. These efforts omitted public involvement, social equity analysis, and environmental or racial considerations. They dehumanized people and led to displacement, oppression, and loss of community. Architects had similar issues: often their buildings ignored the local context and even the needs of users in the pursuit of abstract form. Models that looked good on a studio drafting table often didn't work on the ground for building users or neighbors.

The imperative for planners and designers in this century is to do better: to craft proposals that better meet the needs of particular contexts. Professionals need to ask where ideas are coming from and what assumptions underlie them. They need to be willing to admit that they do not have all the answers and that their proposals might not work. They need to be open to other possibilities than the ones they started with. A blend of careful listening, collaboration, learning, and constructive reframing of solutions is often necessary. Humility and empathy are key.

Develop Multiple Leadership Skills

Training individuals to be leaders is a goal of many educational institutions, but processes for doing this are often not well developed. The public tends to think of leaders as extroverted public speakers or media stars who can attract attention, gain followers, and mobilize people. In reality relatively few of us are cut out to play such roles. But there are dozens of other forms of creative leadership that are equally important. Leadership roles include facilitating, mediating, organizing, coordinating, notetaking, reminding, listening, synthesizing, inspiring, and envisioning. All of us can learn at least some of these skills. In a given meeting, the simple act of keeping good notes and reminding people of what was decided last time can be a form of leadership.

True leadership often means letting other people (especially from underrepresented groups) take the lead and making space for them. For planners, it may mean helping elected officials, department heads, and organization directors claim ownership of one's own ideas. Showing others how they can both get significant changes done and also put themselves in a better light for reelection or promotion often accelerates sustainability action. Care must be taken not to just try to keep officials happy, though. Superiors need to be pushed at times too. There are often diplomatic ways to move new ideas into their field of view and point out their merits. If employees are well prepared, well informed, and sensitive to political and budgetary realities, these creative suggestions are often taken seriously. Disadvantages of BAU can be highlighted as well.

Traditional forms of leadership—in which white men with large egos often sought to stamp their brand on the world—created lots of problems. Monuments to many of these individuals are being torn down because of the role they played in systematic oppression, colonialism, racism, and sexism. Rather than being the product of one or two big names, the sustainable city of the future is likely to be the result of collaborative problem-solving among billions of people. We can all reimagine cities across the world and work together to make them happen. The sooner we begin thinking of social change this way, the better.

Run Effective Meetings

For better or worse, most work in the professional world involves meetings. Getting productive results out of these gatherings, whether in person or online, and getting colleagues excited about sustainability endeavors are important skills. They can be practiced and learned.

Effective meeting facilitation can take many forms, depending on the strengths and personality of the facilitator. Being well organized is usually helpful. This can mean establishing a good agenda, building on the accomplishments of previous meetings, and keeping the meeting moving so as to meet the agreed-upon time frame. Making sure everyone is heard and all viewpoints are considered while avoiding talking too much oneself are key skills. Active listening is critical to be able to hear people's concerns, separate their interests from their positions, depersonalize conflict, and come up with creative solutions that address people's concerns.[12]

Repeatedly calling people's attention to the goals of the meeting—and how they contribute to broader sustainability needs of the community, world, or organization—can help cut short digressions and maintain the group's focus. Keeping clear and transparent notes recording points of view and agreements is a further skill. Humor and a light touch help avoid unnecessary posturing by attendees. Celebrating accomplishments and points of agreement can keep the meeting upbeat and positive.

It's almost never good to present proposals to the group as a finished product, since that invites pushback. Rather, keeping the emphasis on collaboration and iterative development of ideas makes people feel heard and allows for creative input. Recording and restating criticisms also helps people feel they've been heard and understood. Conversely, audiences are quick to sense defensiveness and resistance to their ideas, which leads to more pushback.

Setting ground rules at the start of a meeting often avoids problems later on: for example, "No one person dominates the discussion," "Everyone gets heard," "Everyone has valid points of view," "We will respect everyone's time and finish expeditiously," and "We will focus on actually getting something done." Such parameters can help people feel better about spending their time in a meeting, something many of us resist doing.

Develop Sustainability Education

In line with social learning theory, social change can be seen as an educational process. Individuals can structure learning opportunities both for themselves and for their communities. They can continually look for new information on innovative policies, best practices, and creative solutions to problems. They can challenge their own assumptions. They can look for classes, workshops, or webinars that develop skills and knowledge. They can share strategies with peers and be active within professional associations.

For the public, they can highlight innovations, point out new information, underscore points of agreement, articulate lessons learned, and identify topics on which further information is needed. Even if a meeting has resulted in acrimony, much learning may still have occurred, and that can be highlighted. Articulating points of agreement can constitute learning. People may then see opportunities for collaboration and creative thinking.

It is particularly important to demystify policy options and processes for the public. This can be done in part by making sure that everyone understands institutions, players, acronyms, and procedures. Well-presented maps, diagrams, photos, data, and site plans can help communicate information clearly. Presentations can highlight analysis that is important to the topic at hand and avoid wasting time on tangential elements. They can lay out different theoretical ways of understanding situations and the pros and cons of different strategies. It is also important to admit when we have been wrong, inviting constructive criticism. Helping each other understand that there are in fact alternatives to BAU can be accomplished by presenting creative projects that have been done elsewhere, bringing in outside experts, showing visuals of alternative development strategies or trajectories, and having community members themselves brainstorm alternative futures.

Improve Implementation

Implementation skills are often fairly different from those used by visionaries, reformers, organizers, or other types of professionals. Patience is often essential, as well as attention to detail, follow-through, the ability to coordinate between people, and the skill of understanding and demystifying complex processes. One needs to track the initiative day in and day out, realistically evaluating how things are going and making adjustments as needed. Stakeholders need to be updated, successes celebrated, individual contributions honored, and setbacks acknowledged, realistically evaluated, and moved beyond.

One of the consummate implementers of recent decades has been Mary Nichols, chair of the California Air Resources Board (CARB) and other agencies for sixteen years under multiple governors. As a young environmental lawyer in the 1970s she sued the US Environmental Protection Agency to make it require California to clean up air pollution in the Los Angeles Basin. Democratic governor Jerry Brown then appointed her to the CARB board, and through methodical, science-based policy development she built the agency into a powerful implementer of California's increasingly stringent air pollution laws.

In 2007, after service in other agencies and the Clinton administration, Republican governor Arnold Schwarzenegger appointed Nichols chair of CARB for the second time. The agency was then mandated to implement California's new climate mitigation plan, known as A.B. 32. Under Nichols's leadership, CARB turned general legislative goals such as decreasing emissions to 1990 levels by 2020 into forty-two initial actions, while quantifying likely emissions reductions from each. Her endorsement of a cap-and-trade system to reduce emissions from industry later received pushback from environmental justice advocates who did not like it that industries in polluted communities were allowed to keep operating, but it may have been the only option politically at the time. At times Nichols had to go toe-to-toe with Schwarzenegger to ensure that key pieces of the implementation machinery were put in place. Year after year, the agency monitored results and fine-tuned its projections. The 2020 target was met, and the process continued with the legislature setting new 2030 goals. Nichols was never a charismatic public speaker or media personality. Rather, she was a consummate implementer who built a strong agency that could effectively implement regulation on local air pollution and GHG emissions. Largely because of her, California is a global leader in both regards.

Take Care of Ourselves

Let's face it, community politics is difficult. The type of confrontational, late-night city council meeting described in the previous chapter is all too common. Working within public sector bureaucracies, private sector companies, and NGOs can also be challenging. The institutional changes and public engagement strategies discussed in the last chapter can help improve these contexts in the long term. But how do we survive in the meanwhile as individuals, continue our leadership, and figure out ways to be professionally and personally fulfilled?

First of all, we need to be prepared for what we're going to find. Social change is not easy. We need to anticipate the difficult moments and figure out strategies in advance for dealing with them. Everyone has their own strategies for dealing with adversity. But humor helps, and not being defensive but instead hearing constructive criticism and learning from it. Each of us is a product of our race, class, age, gender, sexual orientation, education, family status, and work environment. It is important to acknowledge our positionality and the strengths and weaknesses that we bring to our work. Having good and diverse networks of friends and colleagues also helps, and talking honestly and openly to others.

We must know ourselves. All of us have particular skills and personalities that work better in some contexts than others. Some of us are introverts, some of us are extroverts. Some of us have greater patience than others. Some of us are specialists and like to zero in on some technical or creative enterprise. Others are generalists and like variety. Some of us are people people; others aren't. Some of us do well within large institutions and like that security. Others are best working independently or with a small group of trusted colleagues. Some of us are by nature activists, educators, facilitators, organizers, writers, public speakers, and many other things. Taking a personality test like the Myers-Briggs can be a useful starting point for personal information. A good therapist, career counselor, or mentor can also help us figure out the best fit between our own characteristics, various professional niches, and our hopes for professional fulfillment.

It helps to frame our own work for ourselves in terms of opportunities and successes rather than problems and overwhelming challenges. Yes, those may exist, but let's also give ourselves credit for everything we accomplish. Celebrate every little step forward, even if it's just a constructive meeting in which everyone listened well. Think about our role, however big or small, in that process. Be clear with everyone that we're in a long-haul process of social change. Appreciate people.

In difficult situations, let people know you're human too. You deserve some respect. You'll treat people with respect and listen carefully to them, but you want to be treated that way yourself. You're not just a mindless arm of the neoliberal state, a tool of a capitalist corporation, or an impractical idealist working for an NGO. You're a living, breathing person who is aware of many of the contradictions around you and is trying to do the best you can. You are an agent of change in an imperfect world. In essence, we are all "muddling through."

Finally, keep a balanced life and don't work 24/7. These things aren't always easy, of course. But having a personal life, time with friends and family, downtime, exercise, and time for sleeping is important for most people's happiness. If you're not getting these things, tell people what you really need. Figure out strategies to get those things as much as possible.

LONG-RUN WAYS TO HELP SOCIAL ECOLOGIES EVOLVE

While the preceding tactics may help with day-to-day sustainability work, the more long-term, strategic need is to reimagine and then build healthy and just social ecologies. Perhaps the first step is to deepen our awareness about the structure of societies and to understand influences that drive long-term change either in a specific city or in the society generally. Who holds power in a particular community? What economic, institutional, social, environmental, and technological factors have led to its evolution? How might these change in constructive directions in the future? What is my personal relationship to that power? How do I want to engage with the power system?

Such analysis leads into reimagining. What are alternative scenarios for the city's future? How might those help meet sustainability and equity goals? Can we outline alternatives either in our minds or more formally with other stakeholders through planning processes? How might they help the city's social ecology evolve in healthy directions? If possible, specific policy initiatives coming from these visions should be shaped in such a way that they will survive short-term changes in politics. The next mayor, governor, or president shouldn't be able to sweep away years of painstaking effort.

Institutional reforms will likely be necessary for many future policy directions. For example, local jurisdictions in a metropolitan area might need to collaborate more closely and might do so informally or ask higher levels of government for funding, technical assistance, and/or legislation to create a stronger regional planning agency. Or the tax system might need to be revised to more fairly distribute revenue and reduce incentives for inappropriate or segregated land use planning. Or electoral processes might need to be changed to protect the marginalized and ensure their representation.

Coalition building is fundamental to building political strength behind particular initiatives. Constellations of organizations representing many citizens and types of constituencies are essential to counter institutional inertia or the power of capital. These constellations can be overt, as when organizers formally gather unions, environmental nonprofits, social service organizations, religious denominations, and other constituencies into a new umbrella organization with its own letterhead and decision-making structure. Or they can be informal, based on low-key coordination and networking between different players. Different tasks can be parceled out to different participants. Some organizations might take the lead on media strategy, for example, working to spread certain messages about the issue in question. Others might take the lead on lobbying and political organizing. Others might focus on research, or litigation, or local organizing.

Strengthening civil society is an important long-term strategy for sustainability purposes. The growth of nonprofit organizations worldwide has been one of the most hopeful trends during the last century. Continued expansion is essential to nudge public and private sectors toward stronger and more effective action and to gradually replace parts of the private sector if capitalism is reined in. NGOs shouldn't be expected to provide basic services long-term that the public sector should provide, such as care of the homeless, because their efforts are likely to be too scattered and inconsistent. However, they can play a short-term role in addressing emergencies and longer-term roles in certain parts of the economy. For example, the housing sector might function best if coordinated among public agencies, private developers, and nonprofit cooperatives, land trusts, cohousing communities, and community development corporations. The latter institutions of civil society can often operate more flexibly, quickly, and creatively than government and often have a greater commitment to sustainability objectives than the private sector.

Professionals can take specific steps at many levels to support the growth of civil society. Government staff can cultivate connections with the staff of nonprofits and can ensure that these organizations are notified and involved in planning and program-building meetings. Public sector contracts can be given preferentially to nonprofits (and to businesses owned by people of color and women), for example having nonprofit health care providers rather than for-profits provide health benefits to public sector workers. Grants, loans, and technical assistance can be channeled to NGOs. At a much larger scale, international development assistance can be preferentially routed through nonprofit organizations rather than for-profit contractors.

Building the diversity of individuals and organizations within social ecologies is a further way to strengthen these systems in the long run. Although homogenous societies may have an easier time of developing consensus around actions, building diversity is critical to equity and justice, and has many long-term advantages. Ideally, all members of a diverse society will learn tolerance, benefit from a wide range of viewpoints and skills, and develop values of empathy and compassion for others different from themselves. As a result they will likely be more creative at problem-solving.

Professionals can help the process along. They can proactively invite diverse people and groups to meetings and can ensure that an atmosphere is established in which everyone feels comfortable participating. They can prevent certain players from dominating discussions and can establish ground rules that everyone has an equal voice. They can promote values of tolerance and equity through messaging and educational campaigns. They can spearhead policy changes to empower the disadvantaged.

Societies that get stuck in conventional ways of doing things often have difficulty incorporating new ideas, experimenting, or adapting to changed circumstances. The result can be the phenomenon that historian Barbara Tuchman identified as "wooden-headedness," in which new ideas are unable to penetrate existing mindsets.[13] Creative, outside-the-box forms of analysis and expression can help counteract this rigidity. Good leadership can solicit and encourage creativity and isn't threatened by new ideas.

CONCLUSION

How much can each of us really do to bring about more sustainable and equitable communities, given all the difficulties of politics, economics, structural racism and sexism, and recalcitrant institutions? Potentially quite a lot. But the process may not be easy or smooth, and some situations are easier to work in than others. Figuring out good niches for ourselves, setting achievable goals within those positions, and developing skills related to analysis, teamwork, leadership, facilitation, and communication are all essential.

There will be times to take stands on principle and times to walk away. There will be times to organize inside existing institutions, and times to take to the streets. Sustainability professionals need to learn to wear many hats. They also need backbone and an ethical compass. Sometimes they will need to say no and to communicate the reasons for their decision so that others can learn from it. Architects at times need to avoid taking on projects if, say, an expensive house would be located in a natural area where no house should really be built. Planners may need to refuse to approve an economic development plan that would increase sprawl, gentrification and displacement, GHG emissions, or dependency on out-of-town corporations. Public agency staff people may need to resist policies of a populist government bent on dismantling environmental protections or else leave their positions to organize resistance outside government. Remaining in jobs in which we lead communities away from a sustainable and equitable future is both ethically suspect and personally deflating.

Bringing about social change, for sustainable cities or any other goal, is not easy. But it is possible to strategically build our own knowledge and skills and then to seek out situations where progress is possible. Some positive change is possible almost everywhere. The challenge is to look for these constructive opportunities and to do everything possible to maximize our impact once within them.

Conclusion

In this book we have asked you to reimagine sustainable cities. We have done this through a tour of big-picture questions about the future. Our goal has been to start a conversation about sustainability and equity that we hope you will continue. How can cities and societies become more just? Address structural racism? Defuse the climate crisis? Protect and restore urban ecosystems? Develop more sustainable economies? Provide everyone with opportunities to live whole, fulfilling, and healthy lives?

Reimagining cities means admitting that BAU is not working—that the systems we live in are unjust and not sustainable. Anyone who has seen the climate projections of the future, or the way people of color and the poor are treated in many countries, or the inadequate ways that democracy is working in many places, knows that there is no choice but to acknowledge the fundamental dysfunction of many current systems and imagine a new reality.

There are watershed moments when societies can choose different futures. This is one now. The Covid-19 pandemic and the global antiracism movement have forced people in many countries to recognize the structural problems that plague our social, economic, and political systems. Collectively, we can embrace the idea of imagining something better and taking action.

Moving toward sustainable cities and societies requires more long-term, holistic, and action-oriented ways of thinking than used in the past. All of us are in the early stages of learning them and understanding the roles we can take. It is important to avoid rigid ideologies and to understand the dynamic, intersectional, and complex systems around us. Then comes the challenge of working to change these systems.

The theme of social ecology is central to this book. For long-lasting change to come about, institutions, economies, political systems, cultures, and individual mindsets must coevolve so as to support constructive action around the challenges of our time. A society in which Fox News is the main information source will never do this. Countries in which for-profit corporations and the wealthy are allowed to run government are unlikely to do this as well, and will be likely to value the self-interest of elites above equality or sustainability. New influences on social evolution must be pioneered instead so that societies move toward healthier social ecologies capable of inclusively planning for sustainability and justice.

We thank you, our readers, for following us this far. But now we ask you to further explore sustainable and just cities on your own. There is no reason why the world of 2100, or perhaps even 2050 or 2030, shouldn't feature carbon-neutral societies with affordable housing, racial justice, women's equality, social tolerance, LGBTQ rights, sustainable economies, healthy and participatory governance, and fun and nurturing public spaces. There is no reason why every child born shouldn't have opportunities to thrive and be happy. With all the wealth, talent, technology, and knowledge in the world today, these things are very possible. So let us now challenge ourselves and each other to reimagine our communities and bring about change.

Acknowledgments

This book more than most is a work of synthesis, drawing upon innumerable sources and influences from throughout our professional and personal lives. For all those contributions, some based on workplace relationships and some not, we are deeply grateful.

We are also particularly grateful to multiple cohorts of undergraduate students at the University of California, Davis, and Temple University who have read chapter drafts for classes and given us feedback and who have helped shape our thinking about these issues. We are particularly grateful to the following students and colleagues for their useful comments and assistance: Sudikshya Bhandari, Bethany Celio, Nermin Dessouky, Christopher N. Jones, Richard LeGates, Gulnara Nabiyeva, Raiza Pilatowski, Darrell Slotton, and Kang-li Wu. We also wish to acknowledge two anonymous reviewers who supplied excellent comments and suggestions on a previous draft. Stacy Eisenstark at UC Press supplied excellent advice and encouragement at numerous stages of the process, assisted by Enrique Ochoa-Kaup and Naja Pulliam Collins. Elisabeth Magnus copyedited the manuscript with admirable professionalism. Claudia Smelser and Nicole Hayward produced a strong design. Emilia Thiuri served ably as production editor, Holly Knowles prepared an excellent index, and Teresa Iafolla and David Olsen managed marketing and publicity. Bill McKibben, Julian Agyeman, Karen Seto, and Richard LeGates generously provided endorsements of the book, for which we are very grateful. Many thanks to all.

Steve Wheeler would particularly like to thank Timothy Beatley, David R. Brower, Manuel Castells, Allan Jacobs, Donella Meadows, Carolyn Merchant, and Richard Norgaard for their past mentorship, friendship, and inspiration to think in terms of changing the world. He would also like to thank the authors Julian Agyeman, Thomas Berry, Octavia Butler, Earnest Callenbach, Fritjof Capra, David Graeber, David Harvey, Paul Krugman, Ursula K. LeGuin, Joanna Macy, Bill McKibben, Michael Moore, Eleanor Ostrom, Kim Stanley Robinson, Rebecca Solnit, Ann Whiston Spirn, and Astra Taylor, among so many others, for their development of visions of a more sustainable and just world in the future. On a more personal level, he gives deepest thanks to his wife Mimi Kusch for her support, love, and insightful editing of several drafts, and to his daughter Madison Rose Kusch Wheeler for the love and joy she has brought to their lives.

Tina Rosan would like to thank Temple University and the College of Liberal Arts for a sabbatical in Spring 2020 and would like to thank Melissa Gilbert, friends, colleagues, and students in the Department of Geography and Urban Studies, as well as Meriel Tulante, for their support. She would also like to give a much-deserved pandemic thanks to her extended family, including Nancy D. Rosan, Richard M. Rosan, Jere Lucey, Richard and Virginia Munkelwitz, Peter and Meg Rosan, Liz and Pete Kirkwood, Caroline and Brian Coffay, and Wendy Costa. And, of course, much thanks and love to her amazing kids Owen, Anya, and Eli, and husband Karl Munkelwitz.

Notes

INTRODUCTION

1. "Sustainability" is usually seen as including social equity goals, just as it also includes economic and environmental dimensions of change. However, since the social equity needs of our time are so great, and have been underemphasized for so long, in this book we often use the phrase "sustainability and equity" to underline them.

2. Donella H. Meadows, Dennis L. Meadows, Jørgen Randers, and William W. Behrens III, *The Limits to Growth: A Report for the Club of Rome on the Predicament of Mankind* (New York: Universe Books, 1972).

3. "Planet of the Year: Endangered Earth," *Time* magazine, January 2, 1989.

4. PBL Netherlands Environmental Assessment Agency, "Trends in Global CO_2 and Total Greenhouse Gas Emissions; 2019 Report," 2020, www.pbl.nl /en/publications/trends-in-global-co2-and-total-greenhouse-gas-emissions-2019 -report.

5. Facundo Alvaredo, Lucas Chancel, Thomas Piketty, Emmanuel Saez, and, Gabriel Zucman, "World Inequality Report 2018," World Inequality Lab, 2018, https://wir2018.wid.world/files/download/wir2018-summary-english.pdf.

6. World Wildlife Fund, "How Many Species Are We Losing?," 2020, https://wwf.panda.org//discover/our_focus/biodiversity/biodiversity/.

7. National Oceanic and Atmospheric Administration, "Ocean Acidification," updated April 2020, www.noaa.gov/education/resource-collections/ocean -coasts-education-resources/ocean-acidification.

8. Bill McKibben, *Falter: Has the Human Game Begun to Play Itself Out?* (New York: Holt, 2019), 1.

9. According to the World Food Programme, the number of hungry people declined by 216 million between 1990–92 and 2020, despite the world's

population growing by 1.9 billion. See World Food Programme, "Zero Hunger," 2020, www.wfp.org/zero-hunger.

10. Anthony Cilluffo and Neil G. Ruiz, "World's Population Is Projected to Nearly Stop Growing by the End of the Century," Pew Research, Fact Tank, June 17, 2019, www.pewresearch.org/fact-tank/2019/06/17/worlds-population-is-projected-to-nearly-stop-growing-by-the-end-of-the-century/.

11. United Nations, "Take Action for the Sustainable Development Goals," 2015, www.un.org/sustainabledevelopment/sustainable-development-goals/.

12. United Nations, "The New Urban Agenda: Key Commitments," *Sustainable Development Goals* (blog), October 20, 2016, www.un.org/sustainable development/blog/2016/10/newurbanagenda/.

13. See, for example, Richard Norgaard, *Development Betrayed: The End of Progress and a Coevolutionary Revisioning of the Future* (London: Routledge, 1996); Stephen M. Wheeler, *Climate Change and Social Ecology* (London: Routledge, 2012).

14. Charles Darwin, *The Descent of Man* (London: John Murray, 1871), http://darwin-online.org.uk/EditorialIntroductions/Freeman_TheDescentof-Man.html; Karl Marx, *Capital*, vol. 1 (1867; repr., Moscow: Progress, 1887), www.marxists.org/archive/marx/works/download/pdf/Capital-Volume-I.pdf; Herbert Spencer, *Principles of Biology* (1864; repr., Washington, DC: Ross and Perry, 2002).

15. For the Chicago School, see Roderick McKenzie, Robert Park, and Ernest Burgess, *The City* (1925; repr., Chicago: University of Chicago Press, 1967). For Murray Bookchin's work, see *Post-scarcity Anarchism* (Berkeley, CA: Ramparts Press, 1971) and *The Ecology of Freedom* (Palo Alto, CA: Cheshire Books, 1982). For more recent writers, see Meadows et al., *Limits to Growth;* Norgaard, *Development Betrayed;* Stewart Brand's work, exemplified in issues of the *Coevolution Quarterly* that he edited between 1974 and 1985 plus his wide range of fascinating books on specific topics; and Eleanor Ostrom, *Governing the Commons: The Evolution of Institutions for Collective Action* (Cambridge: Cambridge University Press, 1990). Other resources on coevolutionary thought include Erich Jantsch, *The Self-Organizing Universe: Scientific and Human Implications of the Emerging Paradigm of Evolution* (New York: Pergamon Press, 1980); Joanna Macy, *World as Lover, World as Self* (Berkeley, CA: Parallax Press, 1991); Edward Goldsmith, *The Way: An Ecological World View* (Boston: Shambhala, 1992); Ervin Laszlo, *The Systems View of the World: A Holistic Vision for Our Time* (Cresskill, NJ: Hampton Press, 1996); and Thomas Homer-Dixon, *The Upside of Down: Catastrophe, Creativity, and the Renewal of Civilization* (Washington, DC: Island Press, 2006).

16. See, for example, Thich Nhat Hanh, *Peace Is Every Step: The Path of Mindfulness in Daily Life* (New York: Bantam, 1991); H.H. the Dalai Lama, *Beyond Religion: Ethics for a Whole World* (Boston: Houghton Mifflin Harcourt, 2020); Helena Norberg-Hodge, *Ancient Futures*, 3rd ed. (White River Junction, VT: Local Futures, 2016); Thomas Berry, *The Great Work: Our Way into the Future* (New York: Bell Tower, 1999); Jerry Mander, *In the Absence of the Sacred: The Failure of Technology and the Survival of the Indian Nations* (San Francisco: Sierra Club Books, 1991).

17. This change, of course, has been under way for a while, with many late twentieth-century authors trumpeting it. See, for example, Gregory Bateson, *Steps to an Ecology of Mind: Collected Essays in Anthropology, Psychiatry, Evolution, and Epistemology* (Chicago: University of Chicago Press, 1972); Frijof Capra, *The Turning Point: Science, Society, and the Rising Culture* (New York: Simon and Schuster, 1982); and John Seed, Joanna Macy, Pat Fleming, and Arne Naess, *Thinking Like a Mountain: Toward a Council of All Beings* (Gabriola Island, BC: New Society, 1988). Many have also argued that indigenous societies throughout history have perceived the world in more holistic ways than twentieth-century thinkers. See, for example, Jerry Mander, *In the Absence of the Sacred: The Failure of Technology and the Survival of the Indian Nations* (San Francisco: Sierra Club Books, 1992).

18. United Nations Population Division, *World Urbanization Prospects: The 2018 Revision,* Department of Economic and Social Affairs, 2019, https://population.un.org/wup/Publications/Files/WUP2018-Report.pdf.

19. C40 Cities, *Why Cities: Ending Climate Change Begins in the City,* 2012, www.c40.org/ending-climate-change-begins-in-the-city.

20. United Nations Centre for Human Settlements (HABITAT), "Hot Cities: Battle-Ground for Climate Change," Nairobi, Kenya, 2020, http://mirror.unhabitat.org/downloads/docs/E_Hot_Cities.pdf.

21. Max Galka, "What Does New York Do with All Its Trash? One City's Waste—in Numbers," *The Guardian,* October 27, 2016.

1. HOW DO WE GET TO CLIMATE NEUTRALITY?

1. Jason Samenow, "Africa May Have Witnessed Its All-Time Hottest Temperature Thursday: 124 Degrees in Algeria," *Washington Post,* July 6, 2018.

2. Lefteris Karagiannopoulos, "In Hot Water: How Summer Heat Has Hit Nordic Nuclear Power Plants," Reuters, August 1, 2018.

3. Mujib Mashal, "India Heat Wave, Soaring up to 123 Degrees, Has Killed at Least 36," *New York Times,* June 13, 2019.

4. Carolyn Gramling, "A Siberian Town Hit 100 Degrees, Setting a New Record for the Arctic Circle," *ScienceNews,* June 23, 2020.

5. Concepción De Leon and John Schwartz, "Death Valley Just Recorded the Hottest Temperature on Earth," *New York Times,* August 17, 2020.

6. We will generally use the terms *global warming* and *climate crisis* in this book rather than "climate change." The former are more accurate and descriptive, whereas in the 1990s climate denier organizations settled on the latter term as the least likely to worry people and lead to calls for action.

7. "Climate and Covid: Converging Crises," *The Lancet,* December 2, 2020, www.thelancet.com/journals/lancet/article/PIIS0140–6736(20)32579–4/fulltext.

8. S. Sherwood, M. J. Webb, J. D. Annan, K. C. Armour, P. M. Forster, J. C. Hargreaves, G. Hegerl, et al., "An Assessment of Earth's Climate Sensitivity Using Multiple Lines of Evidence," *Reviews of Geophysics* 58, no. 4 (December 2020), https://doi.org/10.1029/2019RG000678.

9. Isabel Hilton, "The Reality of Global Warming: Catastrophes Dimly Seen," *World Policy Journal* 25, no. 1 (2008): 1–8.

10. Adil Najam, Saleemul Huq, and Youba Sokona, "Climate Negotiations beyond Kyoto: Developing Countries' Concerns and Interests," *Climate Policy* 3 (2003): 221–31.

11. John Cook, Geoffrey Supran, Stephan Lewandowsky, Naomi Oreskes, and Ed Maibach, *America Misled: How the Fossil Fuel Industry Deliberately Misled Americans about Climate Change* (Fairfax, VA: George Mason University Center for Climate Change Communication, 2019), www.climatechangecommunication.org/wp-content/uploads/2019/10/America_Misled.pdf; Naomi Oreskes and Erik M. Conway, *Merchants of Doubt: How a Handful of Scientists Obscured the Truth on Issues from Tobacco Smoke to Global Warming* (New York: Bloomsbury Press, 2011).

12. City of Vancouver, "Greenest City 2017–18 Implementation Update," 2019, https://vancouver.ca/files/cov/greenest-city-action-plan-implementation-update-2017–2018.pdf; C40 Cities, "Dubai's 'Mohammed Bin Rashid Al Maktoum' 5,000 MW Solar Park Aims to Save 6.5 Million tCO2e Annually," April 15, 2019, www.c40.org/case_studies/dubai-s-mohammed-bin-rashid-al-maktoum-5-000mw-solar-park-aims-to-save-6-5-million-tco2e-annually.

13. International Renewable Energy Agency, *Renewable Power Generation Costs in 2019*, June 2020, www.irena.org/publications/2020/Jun/Renewable-Power-Costs-in-2019.

14. Solar Energy Industries Association, "Renewable Energy Standards," accessed March 18, 2021, www.seia.org/initiatives/renewable-energy-standards; National Conference of State Legislatures, "State Renewable Portfolio Standards and Goals," March 9, 2021, www.ncsl.org/research/energy/renewable-portfolio-standards.aspx.

15. Galen Barbose, *U.S. Renewables Portfolio Standards 2019 Annual Status Update* (Berkeley, CA: Lawrence Berkeley National Laboratory, 2019).

16. Adam Vaughan, "EU Raises Renewable Energy Targets to 32% by 2030," *The Guardian,* June 14, 2018.

17. Kerstine Appunn, Freja Eriksen, and Julian Wettengel, "Germany's Greenhouse Gas Emissions and Energy Transition Targets," *Clean Energy Wire,* March 16, 2021, www.cleanenergywire.org/factsheets/germanys-greenhouse-gas-emissions-and-climate-targets.

18. Synapse Energy Economics, Inc., "An Analysis of Municipalization and Related Utility Practices," September 30, 2017, https://doee.dc.gov/sites/default/files/dc/sites/ddoe/publication/attachments/An%20Analysis%20of%20Municipalization%20and%20Related%20Utility%20Practices.pdf. See also American Public Power Association, "Municipalization," 2020, www.publicpower.org/municipalization.

19. American Public Power Association, "Public Power," accessed March 18, 2021, www.publicpower.org/public-power.

20. Power-Technology, "The Ten Biggest Energy Companies in 2018," updated January 28, 2020, www.power-technology.com/features/top-10-power-companies-in-the-world/.

21. Local Power, "It's Not Just Green. It's Local, & It's Yours. Welcome to Climate Mobilization," 2020, www.localpower.com/index.html.

22. Linda Poon, "What Will It Take to Make Buildings Carbon Neutral?," *Bloomberg CityLab*, September 13, 2018, www.bloomberg.com/news/articles /2018-09-13/how-cities-can-get-serious-about-green-building.

23. Fabrizio Ascione, Nicola Bianco, Olaf Böttcher, Robert Kaltenbrunner, and Giuseppe Peter Vanoli, "Net Zero-Energy Buildings in Germany: Design, Model Calibration, and Lessons Learned from a Case-Study in Berlin," *Energy and Buildings* 133 (2018): 688–710.

24. See Passivehouse.com, "What Is a Passive House?," accessed March 18, 2021, https://passivehouse.com/02_informations/01_whatisapassivehouse/01_ whatisapassivehouse.htm.

25. Delia D'Agostino and Danny Parker, "A Framework for the Cost-Optimal Design of Nearly Zero Energy Buildings (NZEBs) in Representative Climates across Europe," *Energy* 149 (2018): 814–29.

26. Sherri Billimoria, Leia Guccione, Mike Henchen, and Leah Louis-Prescott, "The Economics of Electrifying Buildings," Rocky Mountain Institute, report, 2018, https://rmi.org/insight/the-economics-of-electrifying-buildings/.

27. Davis Rogers, "At $2.3 Trillion, Trump Tax Cuts Leave Big Gap," *Politico*, February 28, 2018.

28. Steve Nadel, *Comparative Energy Use of Residential Gas Furnaces and Electric Heat Pumps* (Washington, DC: American Council for an Energy-Efficient Economy, 2016).

29. Small exceptions for heating and other purposes might be allowed.

30. Grischa Perino and Thomas Pioch, "Banning Incandescent Light Bulbs in the Shadow of the EU Emissions Trading Scheme," *Climate Policy* 17, no. 5 (2017): 678–86.

31. William Goetzler, Shalom Goffri, Sam Jasinski, Rebecca Legett, Heather Lisle, Aris Marantan, Matthew Millard, et al., *Energy Savings Potential and R&D Opportunities for Commercial Building Appliances* (Washington, DC: US Department of Energy Building Technologies Office, 2016).

32. Cristoph J. Meinrenken, D. Chen, R. A. Esparza, V. Iyer, S. P. Paridis, A. Prasad, and E. Whillas, "Carbon Emissions Embodied in Product Value Chains and the Role of Life Cycle Assessment in Curbing Them," *Nature Scientific Reports* 10 (2020): 61–84.

33. "California Moves to End Sales of New Gas-Powered Cars and Trucks by 2035," CNBC, September 23, 2020, www.cnbc.com/2020/09/23/california -moves-to-end-sales-of-new-gas-powered-cars-and-trucks-by-2035.html.

34. Alanna Petroff, "These Countries Want to Ditch Gas and Diesel Cars," CNN Business, July 26, 2017, https://money.cnn.com/2017/07/26/autos/countries -that-are-banning-gas-cars-for-electric/index.html.

35. Bengt Halvorson, "GM Battery Chief: 600-Mile EVs Viable, Million-Mile Battery in Sight," *Green Car Reports*, May 20, 2020.

36. Michael Kent, "EV Fast Charging in Transition: What's Currently Installed, Coming Soon and Still Years Away?," *Charged*, May 10, 2017, https:// chargedevs.com/features/ev-fast-charging-in-transition-whats-currently-installed -coming-soon-and-still-years-away/.

37. Mark Lewis, "The Silent Boom—Norway's Electric Car Revolution," *Independent Media*, January 2, 2019, https://chargedevs.com/features/ev-fast

-charging-in-transition-whats-currently-installed-coming-soon-and-still-years
-away/.

38. Environmental and Energy Study Institute, "Fact Sheet: The Growth in Greenhouse Gas Emissions from Commercial Aviation," 2017, www.eesi.org /papers/view/fact-sheet-the-growth-in-greenhouse-gas-emissions-from-commercial -aviation#5.

39. Richard Martin, "The Race for the Ultra-efficient Jet Engine of the Future," *MIT Technology Review,* March 23, 2016.

40. International Civil Aviation Organization, "International Civil Aviation Organization's Carbon Offset and Reduction Scheme for International Aviation (CORSIA)," Policy Update, February 2017, www.theicct.org/sites/default/files /publications/ICAO%20MBM_Policy-Update_13022017_vF.pdf.

41. Grant Martin, "KLM Encouraged Passengers to Reconsider Flying KLM in New Sustainability Campaign," Forbes, July 7, 2019.

42. Robert Muggah, "Lockdowns Have Been Amazing for the Environment, but COVID-19 Won't Heal the Planet," *Foreign Policy,* August 21, 2020, https:// foreignpolicy.com/2020/08/21/pandemic-lockdowns-climate-environment/.

43. World Bank, "Putting a Price on Carbon with a Tax," accessed March 18, 2021, www.worldbank.org/content/dam/Worldbank/document/SDN/background-note_carbon-tax.pdf.

44. World Bank, *State and Trends of Carbon Pricing 2020* (Washington, DC: World Bank, 2020).

45. Caroline Brouillard and Sarah Van Pelt, "A Community Takes Charge: Boulder's Carbon Tax," February 2007, www-static.bouldercolorado.gov/docs /boulders_carbon_tax-1-201701251557.pdf?_ga=2.211436444.826370937 .1559236046-598171339.1559236046.

46. City of Boulder, Colorado, "Cap Tax," 2021, https://bouldercolorado .gov/climate/climate-action-plan-cap-tax.

47. E.g., Resources for the Future, "Carbon Pricing 101," 2019, www.rff .org/publications/explainers/carbon-pricing-101/.

48. Sabine Fuss, Christian Flachsland, Nicolas Koch, Ulrike Kornek, Brigitte Knopf, and Ottmar Edenhofer, "A Framework for Assessing the Performance of Cap-and-Trade Systems: Insights from the European Union Emissions Trading System," *Review of Environmental Economics and Policy* 12, no. 2 (2018): 220–41.

49. Sunrise Movement, "Green New Deal," www.sunrisemovement.org /green-new-deal/.

50. Steven J. Davis and Ken Caldeira, "Consumption-Based Accounting of CO_2 Emissions," *Proceedings of the National Academy of Sciences* 107, no. 12 (2010): 5687–92.

51. World Resources Institute, *Creating a Sustainable Food Future* (Washington, DC: World Resources Institute, 2019); Hannah Ritchie, "Food Production Is Responsible for One-Quarter of the World's Greenhouse Gas Emissions," Our World in Data, 2019, https://ourworldindata.org/food-ghg-emissions.

52. C. Hoolohan, M. Berners-Lee, J. McKinstry-West, and C.N. Hewitt, "Mitigating the Greenhouse Gas Emissions Embodied in Food through Realistic Consumer Choices," *Energy Policy* 63 (2013): 1065–74.

53. Hoolohan et al., "Mitigating the Greenhouse Gas Emissions."

54. Project Drawdown, "Reduced Food Waste," 2020, www.drawdown .org/solutions/reduced-food-waste.

55. Tobenna D. Anekwe and Ilya Rahkovsky, "Economic Costs and Benefits of Healthy Eating," *Current Obesity Reports* 2 (2013): 225–334.

56. E.g., Frances Seymour, "Seeing the Forests as Well as the (Trillion) Trees in Corporate Climate Strategies," *One Earth* 2, no. 5 (2020): 390–93.

57. World Resources Institute, *Creating a Sustainable Food Future.*

58. See, for example, Jon Gertner, "The Tiny Swiss Company That Thinks It Can Help Stop Climate Change," *New York Times Magazine,* February 12, 2019.

59. Anders Thorbjörnsson, Henrik Wachtmeister, Jianliang Wang, and Mikael Höök, "Carbon Capture and Coal Consumption: Implications of Energy Penalties and Large Scale Deployment," *Energy Strategy Reviews* 7 (2015): 18–28.

60. Daniel L. Sanchez, James H. Nelson, Josiah Johnston, Ana Mileva, and Daniel M. Kammen, "Biomass Enables the Transition to a Carbon-Negative Power System across Western North America," *Nature Climate Change* 5, no. 3 (2015): 230–34.

61. See, for example, David C. Marvin, Dick Cameron, Erik Nelson, Andrew Plantinga, Justin Breck, Gokce Sencan, and Michelle Passero, *Toward a Carbon Neutral California: Economic and Climate Benefits of Land Use Interventions* (San Francisco: Next 10, 2018), http://next10.org/sites/default/files/toward -carbon-neutral-california-web.pdf.

62. Naomi E. Vaughan and Timothy M. Lenton, "A Review of Geoengineer-ing Proposals," *Climatic Change* 109, nos. 3–4 (2011): 745–90.

63. Michael C. MacCracken, "On the Possible Use of Geoengineering to Moderate Specific Climate Change Impacts," *Environmental Research Letters* 4, no. 4 (October–December 2009), doi: 10.1088/1748–9326/4/4/045107.

64. Paul Hawken, ed., *Drawdown: The Most Comprehensive Plan Ever Pro-posed to Reverse Global Warming* (New York: Penguin Books, 2017).

65. E.g., Matthew McKinzie, "NRDC Analysis: Nuclear Energy and a Safer Climate Future," Natural Resources Defense Council, September 29, 2017, www.nrdc.org/experts/matthew-mckinzie/nrdc-analysis-nuclear-energy-and -safer-climate-future.

66. E.g., Stefen Samarripas and Caetano de Campos Lopes, *Taking Stock: Links between Local Policy and Building Energy Use across the United States* (Washington, DC: American Council for an Energy-Efficient Economy, 2020).

67. McKinsey & Company, *Impact of the Financial Crisis on Carbon Eco-nomics,* report, January 1, 2010, www.mckinsey.com/business-functions /sustainability-and-resource-productivity/our-insights/impact-of-the-financial -crisis-on-carbon-economics-version-21.

68. C. J. Polychroniou, "It's Time to Nationalize the Fossil Fuel Industry," *Truthout,* June 26, 2020, https://truthout.org/articles/its-time-to-nationalize-the-fossil-fuel-industry/; Peter Gowan, "A Plan to Nationalize Fossil-Fuel Com-panies," *Jacobin,* March 2018, https://jacobinmag.com/2018/03/nationalize -fossil-fuel-companies-climate-change.

2. HOW DO WE ADAPT TO THE CLIMATE CRISIS?

1. Benjamin P. Horton, J. S. H. Lee, T. A. Shaw, B. P. Horton, N. S. Khan, N. Cahill, A. J. Garner, et al. 2020. "Estimated Global Mean Sea-Level Rise and Its Uncertainties by 2100 and 2300 from an Expert Survey," *Climate and Atmospheric Science* 3, no. 1 (2020), doi: 10.1038/s41612–020–0121–5.

2. City of Fort Lauderdale, "Green Your Routine," accessed March 18, 2021, https://gyr.fortlauderdale.gov/greener-government/climate-resiliency/climate -and-weather-in-fort-lauderdale/the-story-of-sea-level-rise-in-fort-lauderdale /how-do-we-prepare-for-rising-seas.

3. S. A. Kulp and B. H. Strauss, "New Elevation Data Triple Estimates of Global Vulnerability to Sea-Level Rise and Coastal Flooding," *Nature Communications* 10 (2019), article no. 4844.

4. C. Robinson, B. Dilkina, and J. Moreno-Cruz, "Modeling Migration Patterns in the USA under Sea Level Rise," *PLoS ONE* 15, no. 1 (2020), article no. e0227436, doi: 10.1371/journal.pone.0227436.

5. Steve Cohen, "The Politics and Cost of Adapting to Climate Change in New York City," *Phys.org News,* January 21, 2020, https://phys.org/news /2020-01-politics-climate-york-city.html.

6. US General Accounting Office, *Climate Change: Information on Potential Economic Effects Could Help Guide Federal Efforts to Reduce Fiscal Exposure,* GAO 17–720 (Washington, DC: General Accounting Office, 2017).

7. Coral Davenport, "Major Climate Report Describes a Strong Risk of Crisis as Early as 2040," *New York Times,* October 7, 2018.

8. Glenn S. Johnson and Shirley A. Rainey, "Hurricane Katrina: Public Health and Environmental Justice Issues Front and Centered," *Race, Gender and Class* 14, nos. 1/2 (2007): 17–37.

9. Andreas Malm and Rikard Warenius, "The Grand Theft of the Atmosphere: Sketches for a Theory of Climate Injustice in the Anthropocene," in *Climate Futures: Reimagining Global Climate Justice,* ed. Kum-Kum Bhavnani, John Foran, Priya A. Kurian, and Debabish Munshi (London: Zed Books, 2019).

10. Naomi Klein, *This Changes Everything: Capitalism vs. the Climate* (New York: Simon and Schuster, 2014).

11. Julian Agyman, *Introducing Just Sustainabilities: Policy, Planning, and Practice* (London: Zed Books, 2013).

12. Mark Pelling, *Adaptation to Climate Change: From Resilience to Transformation* (New York: Routledge, 2011); W. Neil Adger, Jouni Paavola, Saleemul Huq, and M. J. Mace, eds., *Fairness in Adaptation to Climate Change* (Cambridge, MA: MIT Press, 2006).

13. Data Center, "Lower Ninth Ward Statistical Area," last updated February 24, 2021, www.datacenterresearch.org/data-resources/neighborhood-data /district-8/lower-ninth-ward/.

14. City of Tacoma, "Tacoma Climate Change Resilience Study," May 2016, https://cms.cityoftacoma.org/Sustainability/Climate_Resilience_Study_Final _2016.pdf.

15. City of Boston, "Climate Ready Boston Final Report," 2016, www .boston.gov/departments/environment/preparing-climate-change.

16. Malka Older, "Why Politicians Ignore Disaster Predictions," *Foreign Policy,* April 13, 2020.

17. Rebecca Solnit, *A Paradise Built in Hell: The Extraordinary Communities That Arise in Disaster* (New York: Penguin, 2009).

18. Robert D. Putnam, *Making Democracy Work: Civic Traditions in Modern Italy* (Princeton, NJ: Princeton University Press, 1993).

19. Nermin Dessouky, personal communication, June 2020.

20. City of Boston, "Climate Ready Story Project," accessed March 18, 2021, www.boston.gov/departments/environment/climate-ready-story-project.

21. UK Health and Safety Executive, "Thermal Comfort: The Six Basic Factors," 2020, www.hse.gov.uk/temperature/thermal/factors.htm.

22. Rupa Basu, "High Ambient Temperature and Mortality: A Review of Epidemiologic Studies from 2001 to 2008," *Environmental Health* 8 (2009), doi: 10.1186/1476-069X-8-40.

23. G. Brooke Anderson and Michelle L. Bell, "Heat Waves in the United States: Mortality Risk during Heat Waves and Effect Modification by Heat Wave Characteristics in 43 U.S. Communities," *Environmental Health Perspectives* 119 (2011), article no. 2.

24. F. Canoui-Poitrine, E. Cadot, and A. Spira, "Excess Deaths during the August 2003 Heat Wave in Paris, France," *Epidemiological Review of Public Health* 54, no. 2 (2006): 127–35.

25. "Mortality Increased by 700 during Sweden's Summer Heatwave," *The Local,* December 6, 2018.

26. Stephen M. Wheeler, Yaser Abunnasr, John Dialesandro, Eleni Assaf, Sarine Agopian, and Virginia Carter Gamberini, "Mitigating Urban Heating in Dryland Cities: A Literature Review," *Journal of Planning Literature* 34, no. 4 (2019): 434–46.

27. Wheeler et al., "Mitigating Urban Heating."

28. D. Locke, B. Hall, J. M. Grove, S. T. Pickett, L. A. Ogden, C. Aoki, C. G. Boone, et al., "Residential Housing Segregation and Urban Tree Canopy in 37 US Cities," *SocArcXiv Papers,* January 6, 2020, doi: 10.31235/osf.io/97zcs.

29. Stephanie Pincetl, Thomas Gillespie, Diane E. Pataki, Sassan Saatchi, and Jean-Daniel Saphores, "Urban Tree Planting Programs, Function or Fashion? Los Angeles and Urban Tree Planting Campaigns," *GeoJournal* 78, no. 3 (2013): 475–93.

30. City of Chicago, "Chicago Green Roofs," 2019, www.chicago.gov/city/en/depts/dcd/supp_info/chicago_green_roofs.html.

31. Center for Climate and Energy Solutions, *Resilience Strategies for Extreme Heat,* 2017, www.c2es.org/site/assets/uploads/2017/11/resilience-strategies-for-extreme-heat.pdf.

32. Kibria K. Roman, Timothy O'Brien, Jedediah B. Alvey, and OhJin Woo, "Simulating the Effects of Cool Roof and PCM (Phase Change Materials) Based Roof to Mitigate UHI (Urban Heat Island) in Prominent US Cities," *Energy* 96 (2016): 103–17.

33. Sarah Bretz, Hashem Akbari, and Arthur Rosenfeld, "Practical Issues for Using Solar-Reflective Materials to Mitigate Urban Heat Islands," *Atmospheric Environment* 32, no. 1 (1998): 95–101.

34. Vjolica Berisha, J. R. White, B. McKinney, A. Mohamed, K. Goodin, D. Hondula, M. Roach, et al., "Assessing Adaptation Strategies for Extreme Heat: A Public Health Evaluation of Cooling Centers in Maricopa County, Arizona," *Journal of the American Meteorological Society* 9 (2017): 71–80.

35. Susan M. Bernard and Michael A. McGeehin, "Municipal Heat Wave Response Plans," *American Journal of Public Health* 94, no. 9 (2004): 1520–22.

36. Alexandria Macia, "Council Passes Bill to Require Air Conditioning for Rental Units," February 25, 2020, mymcm.com (Montgomery County, MD), www.mymcmedia.org/council-passes-bill-to-require-air-conditioning-for-rental -housing/.

37. National Hurricane Center, "Costliest U.S. Tropical Cyclones Tables Updated," 2018, www.nhc.noaa.gov/news/UpdatedCostliest.pdf.

38. City of Houston, *Climate Action Plan,* April 22, 2020, http://green houstontx.gov/climateactionplan/.

39. For more information, see New York State's Department of Conservation at www.dec.ny.gov/energy/102559.html.

40. John Lyle, *Regenerative Design for Sustainable Development* (Thousand Oaks, CA: Sage Publications, 1989); Stuart Cowan and Sim Van der Ryn, *Ecological Design* (Washington, DC: Island Press, 1990).

41. Anna Kusmer, "How China's Nature-Based Solutions Help with Extreme Flooding," *The World,* June 31, 2020.

42. US Environmental Protection Agency, "Summary of State Standards," 2011, https://www3.epa.gov/npdes/pubs/sw_state_summary_standards.pdf.

43. City of Washington, DC, "2020 Permeable Surface Rebate Program," 2020, https://doee.dc.gov/service/permeablesurfacerebate.

44. City of Seattle, "Green Stormwater Infrastructure," 2020, www.seattle .gov/utilities/your-services/sewer-and-drainage/green-stormwater-infrastructure.

45. City of Aukland, "Rainwater Collection on Your Property," 2020, www .aucklandcouncil.govt.nz/environment/looking-after-aucklands-water/rainwater -tanks/Pages/rainwater-collection-property.aspx.

46. Philly Watersheds, "Green City, Clean Waters," 2016, http:// phillywatersheds.org/what_were_doing/documents_and_data/cso_long_term _control_plan.

47. M. Heckert and C. D. Rosan, "Developing a Green Infrastructure Equity Index to Promote Equity Planning," *Urban Forestry and Urban Greening* 19 (2016): 263–70.

48. City of Copenhagen, *Copenhagen Climate Adaptation Plan,* 2011, https:// international.kk.dk/sites/international.kk.dk/files/uploaded-files/Copenhagen%20 Climate%20Adaptation%20Plan%20-%202011%20-%20short%20version .pdf; Justin Gerdes, "What Copenhagen Can Teach Cities about Adapting to Climate Change," Forbes, October 31, 2012.

49. Miyuki Hino, Christopher B. Field, and Katharine J. Mach, "Managed Retreat as a Response to Natural Hazard Risk," *Nature Climate Change* 7 (2017): 364–70.

50. Simon A. Andrew, Sudha Arlikatti, Laurie C. Long, and James M. Kendra, "The Effect of Housing Assistance Arrangements on Household Recovery:

An Empirical Test of Donor-Assisted and Owner-Driven Approaches," *Journal of Housing and the Built Environment* 28, no. 1 (2013): 17–34.

51. Umair Irfan, "Scientists Fear the Western Wildfires Could Lead to Long-Term Lung Damage," Vox, September 24, 2020, www.vox.com/21451219/wildfire-2020-california-oregon-washington-health-air-quality.

52. See, for example, California Fire Safe Council, "Fire Safety Information for Residents," 2021, https://cafiresafecouncil.org/resources/fire-safety-information-for-residents/.

53. Dale Kasler, "Valley, Butte Fires among Costliest Ever at $2 Billion in Damages," *Sacramento Bee,* October 14, 2015; Eroc Sagara, "Bad Wiring at House Sparked California's Deadly 2015 Valley Fire," Reveal channel, Center for Investigative Reporting, August 11, 2016.

54. "The 11 Cities Most Likely to Run Out of Drinking Water—Like Cape Town," BBC News, February 11, 2018.

55. T. C. Brown, V. Mahat, and J. A. Ramirez, "Adaptation to Future Water Shortages in the United States Caused by Population Growth and Climate Change," *Earth's Future* 7 (2019): 219–34, doi: 10.1029/2018EF001091.

56. Abrahm Lustgarten, "The Great Climate Migration," *New York Times,* July 23, 2020, www.nytimes.com/interactive/2020/07/23/magazine/climate-migration.html.

57. See, for example, Barb Anderson, "Smart Water Use on Your Farm or Ranch," the section "Water-Conserving Plants," 2021, Sustainable Agriculture Research and Education, www.sare.org/Learning-Center/Bulletins/Smart-Water-Use-on-Your-Farm-or-Ranch/Text-Version/Plant-Management/Water-Conserving-Plants.

58. Daniel A. Gross, "Recycling Sewage into Drinking Water Is No Big Deal. They've Been Doing It in Namibia for 50 Years," *The World,* December 15, 2016, Public Radio International.

59. "Mortality Increased by 700."

60. Intergovernmental Panel on Climate Change (IPCC), "Assessment Reports—Working Group II: Impacts, Adaptation and Vulnerability," 2020, secs. 2.3–2.4, https://archive.ipcc.ch/ipccreports/tar/wg2/index.php?idp=8.

61. Sonja J. Vermeulen, Bruce M. Campbell, and John S. I. Ingram, "Climate Change and Food Systems," *Annual Review of Environment and Resources* 37 (2012): 195–222; IPCC, *Climate Change and Land* (New York: United Nations, 2019), www.ipcc.ch/srccl-report-download-page/.

62. Ashville Buncombe Food Policy Council, "City of Asheville Food Policy Action Plan," 2017, www.abfoodpolicy.org/asheville-buncombe-food-action-plan/.

63. "Ananas New Community: A Food Systems Approach," The Plan, 2016, www.theplan.it/eng/award-2016-urbanplanning/ananas-new-community-a-food-systems-approach-to-habitat-enhancement-and-social-sustainability-philippines-1.

64. James Hansen, M. Sato, P. Hearty, R. Ruedy, M. Kelley, V. Masson-Delmotte, G. Russell, et al., "Ice Melt, Sea Level Rise and Superstorms: Evidence from Paleoclimate Data, Climate Modeling, and Modern Observations

That 2 °C Global Warming Could Be Dangerous," *Atmospheric Chemistry and Physics* 16 (2016): 3761–3812; James Hansen, "Scientific Reticence and Sea Level Rise," *Environmental Research Letters* 2, no. 2 (2007), article no. 024002.

65. Sarah Gibbens, "Hurricane Sandy Explained," *National Geographic,* February 11, 2019, www.nationalgeographic.com/environment/natural-disasters /reference/hurricane-sandy/.

66. PlaNYC, *A Stronger, More Resilient New York* (New York: City of New York, 2013), www1.nyc.gov/site/sirr/report/report.page.

67. Peter G. Peterson Foundation, "U.S. Defense Spending Compared to Other Countries," May 13, 2020, www.pgpf.org/chart-archive/0053_defense -comparison. The US figure in this index does not include Department of Energy and intelligence agency spending, which would take the figure even higher.

68. Christopher Flavelle, "U.S. Flood Strategy Shifts to 'Unavoidable' Relocation of Entire Neighborhoods," *New York Times,* August 26, 2020.

69. Richard Register, *EcoCity Berkeley* (Berkeley, CA: North Atlantic Books, 1988).

70. Montgomery County, Maryland, "Transferable Development Rights," last updated December 4, 2019, https://montgomeryplanning.org/planning /agricultural-reserve/transferable-development-rights/.

3. HOW MIGHT WE CREATE MORE SUSTAINABLE ECONOMIES?

1. Paul Roberts, "This Is What Really Happens When Amazon Comes to Your Town," *Politico,* October 19, 2017.

2. Michael Hobbes, "How Amazon Is Holding Seattle Hostage," *Huffington Post,* May 12, 2018.

3. Jeff Stein, "Seattle Council Votes to Repeal Tax to Help Homeless amid Opposition from Amazon, Other Businesses," *Washington Post,* June 12, 2018.

4. Martin O'Connor, ed., *Is Capitalism Sustainable? Political Economy and the Politics of Ecology* (New York: Guilford Press, 1994).

5. Mary Cappabianca, "Harvard 2016 Spring Poll," Kennedy School Institute of Politics, April 25, 2016, http://iop.harvard.edu/youth-poll/past/harvard-iop -spring-2016-poll.

6. Frank Newport, "Democrats More Positive about Socialism Than Capitalism," Gallup, August 13, 2018, https://news.gallup.com/poll/240725/democrats -positive-socialism-capitalism.aspx.

7. Philip Galanes, "The Mind Meld of Bill Gates and Steven Pinker," *New York Times,* January 27, 2018.

8. Steven Pinker, *Enlightenment Now: The Case for Reason, Science, Humanism, and Progress* (New York: Viking, 2018).

9. Francis Fukuyama, *The End of History and The Last Man* (New York: Free Press, 1992).

10. Francis Fukuyama, *Identity: Contemporary Identity Politics and the Struggle for Recognition* (London: Profile, 2018).

11. For additional thoughts along this line, see Robert B. Reich, *Saving Capitalism: For the Many, Not the Few* (New York: Knopf, 2015).

12. See, for example, Anthony Giddens, *The Third Way: The Renewal of Social Democracy* (Cambridge: Polity Press, 1999); Nik Brandal and Dag Einar Thorsen, *The Nordic Model of Social Democracy* (New York: Palgrave Macmillan, 2013); Hans Keman, *Social Democracy: A Comparative Account of the Left-Wing Party Family* (New York: Routledge, 2017).

13. Judith Westerink, Annet Kempenaar, Marjo van Lierop, Stefan Groot, Arnold van der Valk, and Adri van den Brink, "The Participating Government: Shifting Boundaries in Collaborative Spatial Planning of Urban Regions," *Environment and Planning C: Politics and Space* 35, no. 1 (2016): 147–68.

14. Tony Judt, *Post-war: A History of Europe since 1945* (New York: Penguin Books, 2005).

15. Organisation for Economic Co-operation and Development, "Social Spending Stays at Historically High Levels in Many OECD Countries," Social Expenditure Update, October 2016, www.oecd.org/els/soc/OECD2016-Social-Expenditure-Update.pdf.

16. Anu Partanen, *The Nordic Theory of Everything: In Search of a Better Life* (New York: Harper, 2016); George Lakey, *Viking Economics: How the Scandinavians Got It Right—and How We Can Too* (Brooklyn, NY: Melville House, 2016).

17. William Foote Whyte and Catherine King Whyte, *Making Mondragon: The Growth and Dynamics of the Worker Cooperative Complex,* 2nd ed. (Ithaca, NY: Cornell University Press, 1991).

18. Obery M. Hendricks, "The Uncompromising Anti-capitalism of Martin Luther King, Jr.," *Huffington Post,* updated March 22, 2014, www.huffingtonpost.com/obery-m-hendricks-jr-phd/the-uncompromising-anti-capitalism-of-martin-luther-king-jr_b_4629609.html.

19. Isabella Gomez Sarmiento, "How Evo Morales Made Bolivia a Better Place . . . before He Fled the Country," National Public Radio, November 26, 2019.

20. Michael Lerner, *Revolutionary Love: A Political Manifesto to Heal and Transform the World* (Berkeley: University of California Press, 2019).

21. Kenneth Boulding, "The Economics of the Coming Spaceship Earth," in *Environmental Quality in a Growing Economy,* ed. H. Jarrett (Baltimore: Resources for the Future/Johns Hopkins University Press, 1966), 3–14.

22. See, for example, International Resource Panel Working Group on Decoupling, *Decoupling Natural Resource Use and Environmental Impacts from Economic Growth* (Paris: United Nations Environment Program, 2011), https://wedocs.unep.org/handle/20.500.11822/9816.

23. Paul Hawken, *The Ecology of Commerce: A Declaration of Sustainability* (1993; repr., New York: Harper, 2011); Paul Hawken, *Natural Capitalism: Creating the Next Industrial Revolution* (Boston: Little, Brown, 1999).

24. For a nice discussion of the myth of decoupling, see Tim Jackson, *Prosperity without Growth: Foundations for the Economy of Tomorrow,* 2nd ed. (London: Routledge, 2017).

25. Rick Gladstone, "World Population Could Peak Decades ahead of U.N. Forecast, Study Asserts," *New York Times,* July 14, 2020.

26. Center for the Advancement of the Steady State Economy, "Steady State Economy Definition," 2019, https://steadystate.org/discover/definition/.

27. Herman E. Daly, *Steady-State Economics* (1978; repr., Washington, DC: Island Press, 1991); Herman E. Daly, *From Uneconomic Growth to a Steady-State Economy* (Cheltenham, UK: Edward Elgar, 2014).

28. Herman E. Daly and John Cobb, *For the Common Good* (Boston: Beacon Press, 1989).

29. Michael H. Shuman, *Going Local: Creating Self-Reliant Communities in a Global Age* (New York: Routledge, 2000).

30. See the Alliance for Financial Inclusion at www.afi-global.org.

31. National Commission on the Causes of the Financial and Economic Crisis in the United States, *The Financial Crisis Inquiry Report* (New York: Cosimo, 2011).

32. See, for example, Christopher Mayer, "Housing Bubbles: A Survey," *Annual Review of Economics* 3 (2011): 559–77.

33. Susan Strange, *Casino Capitalism* (Manchester: Manchester University Press, 1986).

34. Jamie Golombek, "Flipping Houses? Expect to Face Tax on 100% of Your Profits," *Financial Post,* February 10, 2016.

35. Grace Blakeley, *Stolen: How to Save the World from Financialisation* (London: Repeater Books, 2019).

36. European Commission, "Enhanced Cooperation on Financial Transaction Tax—Questions and Answers," press release, October 23, 2012, https://ec.europa.eu/commission/presscorner/detail/lt/MEMO_12_799.

37. Alana Semuels, "Loose Tax Laws Aren't Delaware's Fault," *The Atlantic,* October 5, 2016.

38. Leslie Wayne, "How Delaware Thrives as a Corporate Tax Haven," *New York Times,* June 30, 2012.

39. Ralph Nader, Mark J. Green, and Joel Seligman, *Taming the Giant Corporation* (New York: Norton, 1977).

40. Jeanne Sahadi, "California Will Now Require More Diversity on Company Boards," CNN Business, September 30, 2020.

41. Coop Danmark A/S, "About Coop," accessed 2020, https://om.coop.dk/koncern/in+english.aspx.

42. David Thompson, "Japan: Land of Cooperatives," Cooperative Grocer Network, March-April 2008, www.grocer.coop/articles/japan-land-cooperatives.

43. World Council of Credit Unions, "Why Credit Unions?," accessed March 19, 2021, www.woccu.org/impact/credit_unions.

44. Mondragon, "Humanity at Work," accessed 2020, www.mondragon-corporation.com/en/about-us/.

45. Ramon Flecha and Pun Ngai, "The Challenge for Mondragon: Searching for Cooperative Values in Times of Internationalization," *Organization* 21, no. 5 (2014): 666–82.

46. Lakey, *Viking Economics.*

47. C.D. Alexander Evans, "The Future of the Japanese Labor Movement," *Dissent*, August 13, 2011, www.dissentmagazine.org/online_articles/the-future -of-the-japanese-labor-movement.

48. See, for example, Mary Donegan, T. William Lester, and Nichola Lowe, "Striking a Balance: A National Assessment of Economic Development Incentives," Upjohn Institute Working Paper 18–291, W.E. Upjohn Institute for Employment Research, Kalamazoo, MI, https://doi.org/10.17848/wp18–291; Greg LeRoy, *The Great American Jobs Scam: Corporate Tax Dodging and the Myth of Job Creation* (San Francisco: Barrett-Koehler, 2005).

49. Charles Marohn, *Strong Towns: A Bottom-Up Revolution to Rebuild American Prosperity* (Hoboken, NJ: Wiley, 2020).

50. Grace Enda and William Gale, "How Could Changing Capital Gains Taxes Raise More Revenue?," Brookings Institution, January 14, 2020, www .brookings.edu/blog/up-front/2020/01/14/how-could-changing-capital-gains -taxes-raise-more-revenue/.

51. Richard Florida, *The Rise of the Creative Class* (New York: Basic Books, 2002).

52. John P. Kretzmann and John McKnight, *Building Communities from the Inside Out: A Path toward Finding and Mobilizing a Community's Assets* (Evanston, IL: Asset-Based Community Development Institute, 1993).

53. Glenda Cooper and Al Fletcher, "Case Study: Evaluating Hamilton's Neighborhood Action Strategy," Tamarack Institute, accessed 2018, www.tamarack community.ca/library/case-study-evaluating-hamiltons-neighbourhood-action -strategy.

54. Techstars, "Techstars Boulder Accelerator," accessed 2020, www .techstars.com/accelerators/boulder.

55. Rustam Lalkaka, "'Best Practices' in Business Incubation: Lessons (Yet to Be) Learned," European Union International Conference on Business Centers. 2001, http://plan.ystp.ac.ir/documents/14853/2711675/Best%20Practices%20in% 20Business%20Incubation_0.pdf.

56. See San Francisco Office of Economic and Workforce Development, "San Francisco Local Hiring Policy for Construction Fact Sheet," accessed March 19, 2021, https://sfpublicworks.org/sites/default/files/2081-5%20Local%20Hire% 20Fact%20Sheet.pdf.

57. Aditya Chakrabortty, "In 2011 Preston Hit Rock Bottom. Then It Took Back Control," *The Guardian*, January 31, 2018.

58. Elizabeth Strom and Robert Kerstein, "The Homegrown Downtown: Redevelopment in Asheville, North Carolina," *Urban Affairs Review* 53, no. 3 (2015): 495–521.

4. HOW CAN WE MAKE AFFORDABLE, INCLUSIVE, AND EQUITABLE CITIES?

1. Jacki Montgomery, Aila Khan, and Louise Carley Young, "Young Women Share Their Stories of Homelessness," *The Conversation*, June 20, 2019, https:// phys.org/news/2019–06-young-women-stories-homelessness.html.

2. United Way Toronto, "Life in Rexdale's High-Rises," www.homelesshub.ca /resource/life-rexdale's-high-rises; United Way Toronto, *Vertical Poverty: Declining Income, Housing Quality and Community Life in Toronto's Inner Suburban High-Rise Apartments* (Toronto: Phoenix Print, 2011), www.unitedwaygt.org /document.doc?id=89.

3. Manny Fernandez and Audra D. S. Burch, "George Floyd, from 'I Want to Touch the World,' to 'I Can't Breathe,'" *New York Times,* June 18, 2020.

4. Melissa Checker, "Wiped Out by the 'Greenwave': Environmental Gentrification and the Paradoxical Politics of Urban Sustainability," *City and Society* 23, no. 2 (2011): 210–29.

5. Julian Agyeman, Robert D. Bullard, and Bob Evans, *Just Sustainabilities: Development in an Uneven World* (Cambridge, MA: MIT Press, 2005).

6. Andrew T. Young, "How the City Air Made Us Free: The Self-Governing Medieval City and the Bourgeoisie," *Revaluation,* January 17, 2017.

7. Cecilia Tacoli, "Urbanization, Gender, and Poverty," UNFPA Technical Briefing, March 2012, www.unfpa.org/sites/default/files/jahia-publications /documents/publications/2012/UNFPA_gender_March%202012.pdf.

8. Claire Colomb and Mike Raco, "Planning for/in the Super-diverse City: Between Celebratory Policy Narratives and the Reality of Planning Policies in London," paper presented at "The Ideal City: Between Myth and Reality. Representations, Policies, Contradictions and Challenges for Tomorrow's Urban Life," RC21 International Conference, 2015, Urbino, Italy, www.rc21.org/en /wp-content/uploads/2014/12/H3-Colomb_Raco_extended-abstract.pdf.

9. New York City, *State of Our Immigrant City: Annual Report,* March 2018, https://www1.nyc.gov/assets/immigrants/downloads/pdf/moia_annual _report_2018_final.pdf.

10. Suyin Haynes, "As Protesters Shine a Spotlight on Racial Injustice in America, the Reckoning Is Going Global," *Time,* June 11, 2020.

11. United Nations Conference on Human Settlements, "The Vancouver Declaration on Human Settlements," 1976, http://mirror.unhabitat.org/downloads /docs/The_Vancouver_Declaration.pdf.

12. Patrick Sisson, "Solving Affordable Housing: Creative Solutions around the U.S.," *Curbed,* July 25, 2017.

13. Sarah Holder, "For Low-Income Renters, the Affordable Housing Gap Persists," Bloomberg CityLab, March 13, 2018, www.bloomberg.com/news /articles/2018-03-13/low-income-renters-find-stubborn-affordable-housing-gap.

14. May Bulman, "UK Facing Its Biggest Housing Shortfall on Record with Backlog of 4m Homes, Research Shows," *The Independent,* May 18, 2018.

15. Terner Center for Housing Innovation, "Housing in Sweden: An Overview," November 2017, http://ternercenter.berkeley.edu/uploads/Swedish_ Housing_System_Memo.pdf.

16. Maurice Blanc, "The Impact of Social Mix Policies in France," *Housing Studies* 25, no. 2 (2010): 257–72.

17. Feargus O'Sullivan, "The Rent Is Now Somewhat Less High in Paris," Bloomberg CityLab, August 3, 2016, www.citylab.com/equity/2016/08/paris-rent -control-laws-are-working/494282/.

18. John F. Bauman, "Public Housing: The Dreadful Saga of a Durable Policy," *Journal of Planning Literature* 8, no. 4 (1994): 347–61.

19. United Nations Office of the High Commissioner, "Informal Settlements and the Right to Housing," September 19, 2018, www.ohchr.org/EN/Issues /Housing/Pages/InformalSettlementsRighttoHousing.aspx.

20. Watson Institute of International and Public Affairs, Brown University, "Costs of War," accessed December 12, 2018, https://watson.brown.edu /costsofwar/.

21. National Alliance to End Homelessness, *State of Homelessness Report: 2020 Edition* (Washington, DC: National Alliance to End Homelessness, 2020), https://endhomelessness.org/homelessness-in-america/homelessness-statistics /state-of-homelessness-report/.

22. Peter L. D'Antonio, "Soaring House Prices Reflect a Shortage of Homes Rather Than a New Housing Bubble," *Journal of Business and Economic Policy* 7, no. 1 (2020): 28–33.

23. Commonwealth of Massachusetts, "Chapter 40 and MassHousing," accessed 2020, www.masshousing.com/en/programs-outreach/planning-programs /40b.

24. Rachel G. Bratt, "The Quadruple Bottom Line and Nonprofit Housing Organizations in the United States," *Housing Studies* 27, no. 4 (2012): 438–56.

25. Patrick Sisson, "Solving Affordable Housing: Creative Solutions from around the U.S.," *Curbed,* July 25, 2017.

26. Abhas Jha, "'But What about Singapore?' Lessons from the Best Public Housing Program in the World," *Sustainable Cities* (World Bank blog), January 31, 2018, http://blogs.worldbank.org/sustainablecities/what-about-singapore -lessons-best-public-housing-program-world.

27. Jie Chen, Zan Yang, and Ya Ping Wang, "The New Chinese Model of Public Housing: A Step Forward or Backward," *Housing Studies* 29, no. 4 (2014): 534–50.

28. Hannu Ruonavaara, "How Divergent Housing Institutions Evolve: A Comparison of Swedish Tenant Co-operatives and Finnish Shareholders' Housing Companies," *Housing, Theory and Society* 22, no. 4 (2006): 213–36.

29. Dudley Street Neighborhood Initiative, "About Us," accessed 2020, www.dsni.org.

30. Sharon Otterman and Matthew Haag, "Rent Regulations in New York: How They'll Affect Tenants and Landlords," *New York Times,* July 12, 2019.

31. Shereen E. Attia, "Rent Control Dilemma Comeback in Egypt's Governance: A Hedonic Approach," Working Paper 979, Economic Research Forum, Giza; Nermin Dessouky, personal communication, June 2020.

32. See the website of the Eviction Lab, https://evictionlab.org.

33. Matthew Yglesias, "Joe Biden's Surprisingly Visionary Housing Plan, Explained," Vox, July 9, 2020; Center for Budget and Policy Priorities, "Housing Choice Voucher Factsheets," August 9, 2017, www.cbpp.org/housing-choice -voucher-fact-sheets.

34. William G. Gale, "Chipping Away at the Home Mortgage Interest Deduction," *Wall Street Journal,* April 9, 2019.

35. Tony Crook and Sarah Monk, "Planning Gains, Providing Homes," *Housing Studies* 26, nos. 7–8 (2011): 997–1018.

36. Inclusionary Housing, "Where Does Inclusionary Housing Work?," 2019, http://inclusionaryhousing.org/inclusionary-housing-explained/what-is-inclusionary-housing/where-does-it-work-3/.

37. Emily Moss, Kriston McIntosh, Wendy Edelberg, and Kristen E. Broady, "The Black-White Wealth Gap Left Black Households More Vulnerable," *Up Front* (Brookings Institution blog), December 8, 2020, www.brookings.edu/blog/up-front/2020/12/08/the-black-white-wealth-gap-left-black-households-more-vulnerable/.

38. MIT Living Wage Calculator, accessed March 21, 2021, https://livingwage.mit.edu.

39. See, for example, Richard Rothstein, *The Color of Law: A Forgotten History of How Our Government Segregated America* (New York: Liveright, 2017).

40. Adam Nagourney and Jeremy W. Peters, "A Half-Century On, an Unexpected Milestone for L.G.B.T.Q. Rights," *New York Times,* June 15, 2020.

41. Katrin Auspurg, Andreas Schneck, and Thomas Hinz, "Closed Doors Everywhere? A Meta-analysis of Field Experiments on Ethnic Discrimination in Rental Housing Markets," *Journal of Ethnic and Migration Studies* 45, no. 1 (2018): 95–114; Peter Christensen and Christopher Timmins, "Sorting or Steering: Experimental Evidence on the Economic Effects of Housing Discrimination," NBER Working Paper 24826, National Bureau of Economic Research, Washington, DC.

42. Caitlin McCabe, "Wells Fargo to Pay Philly $10 Million to Resolve Lawsuit Alleging Lending Discrimination against Minorities," *Philadelphia Inquirer,* December 16, 2019.

43. NAACP, "Criminal Justice Fact Sheet," 2019, www.naacp.org/criminal-justice-fact-sheet/.

44. See People's Paper Co-op website at http://peoplespaperco-op.weebly.com.

45. Mywage.co.za, "Affirmative Action," accessed March 21, 2021, https://mywage.co.za/decent-work/fair-treatment/affirmative-action.

46. Metropolitan Area Planning Council, "Research: The Diversity Deficit: Municipal Employees in Metro Boston," July 13, 2020, https://metrocommon.mapc.org/reports/14.

47. City of Boston, "Workforce Profile Report" and "Understanding Employee Demographics," accessed March 21, 2021, www.cityofboston.gov/diversity/.

48. Angela Glover Blackwell and Michael McAfee, "Banks Should Face History and Pay Reparations," *New York Times,* June 26, 2020.

49. Annie Lowrey, "A Cheap, Race-Neutral Way to Close the Racial Wealth Gap," *The Atlantic,* June 29, 2020.

50. International Justice Resource Center, "Immigration and Migrants' Rights," Thematic Research Guide, accessed March 21, 2021, https://ijrcenter.org/thematic-research-guides/immigration-migrants-rights/.

51. "How Sweden Treats Refugees," New Matilda, March 26, 2014, https://newmatilda.com/2014/03/26/how-sweden-treats-refugees/.

52. United Nations Educational, Scientific, and Cultural Organization, *Cities Welcoming Refugees and Migrants* (Paris: UNESCO, 2016), www.migration4development.org/sites/default/files/report.pdf.

53. American Planning Association, *Planning for Equity Policy Guide* (Chicago: American Planning Association, 2019, https://planning-org-uploaded-media.s3.amazonaws.com/publication/download_pdf/Planning-for-Equity-Policy-Guide-rev.pdf.

54. Stacy Mitchell, "Tax-Base Sharing—Metropolitan Revenue Distribution, MN," Institute for Local Self-Reliance, accessed March 21, 2021, https://ilsr.org/rule/tax-base-sharing/2301–2/.

55. Tonya Moreno, "States Where Cities and Counties Levy Additional Income Taxes," The Balance (website), updated February 23, 2021, www.thebalance.com/cities-that-levy-income-taxes-3193246.

56. Marc Lee, "The Case for a Progressive Property Tax," *Policynote* (blog), November 2, 2016, www.policynote.ca/the-case-for-a-progressive-property-tax/.

57. Tax Policy Center, "Historical Highest Marginal Income Tax Rates," February 4, 2020, www.taxpolicycenter.org/statistics/historical-highest-marginal-income-tax-rates.

58. National Archives, "Rates of Income Tax," February 7, 2012, https://webarchive.nationalarchives.gov.uk/20120207161704/http://www.hmrc.gov.uk/stats/tax_structure/menu.htm.

59. S. J. Yun, "Personal Income Tax Rates," *Global Finance,* November 18, 2015.

60. Paul Krugman, "Elizabeth Warren Does Teddy Roosevelt," *New York Times,* January 28, 2019, www.nytimes.com/2019/01/28/opinion/elizabeth-warren-tax-plan.html.

61. Kriston McIntosh, Emily Moss, Ryan Nunn, and Jay Shambaugh, "Examining the Black-White Wealth Gap," *Up Front* (Brookings Institution blog), February 27, 2020, www.brookings.edu/blog/up-front/2020/02/27/examining-the-black-white-wealth-gap/.

62. Payscale.com. "CEO Pay: How Much Do CEOs Make Compared to Their Employees?," 2020, www.payscale.com/data-packages/ceo-pay.

63. Philip K. Robins, "A Comparison of the Labor Supply Findings from the Four Negative Income Tax Experiments," *Journal of Human Resources* 20, no. 4 (1985): 567–82.

64. Derek Hum and Wayne Simpson, "A Guaranteed Annual Income? From Mincome to the Millennium," *Policy Options,* January/February 2001.

65. Hilary Hoynes and Jesse Rothstein, "Universal Basic Income in the United States and Advanced Countries," *Annual Review of Economics* 11 (2019): 929–58.

66. William D. Hartung and Mandy Smithberger, "America's Defense Budget Is Bigger Than You Think," *The Nation,* May 7, 2019.

67. Lance Freeman, *There Goes the 'Hood: Views of Gentrification from the Ground Up* (Philadelphia: Temple University Press, 2006); Andres Duany, "Three Cheers for Gentrification," *American Enterprise* 12, no. 3 (2001): 36.

68. In this discussion we are indebted to Ena Lupine, "Inclusionary Sustainability: Anti-displacement Recommendations for California's Climate Investments" (M.S. thesis, University of California, Davis, 2017).

69. Sally C. Curtin and Holly Hedegaard, "Suicide Rates for Females and Males by Race and Ethnicity: United States, 1999 and 2017," National Center for Health Statistics, April 2020, https://save.org/wp-content/uploads/2020/04/rates_1999_2017.pdf.

5. HOW CAN WE REDUCE SPATIAL INEQUALITY?

1. Joint Venture Silicon Valley, *2019 Silicon Valley Index* (San Jose, CA: Joint Venture Silicon Valley, 2019), https://jointventure.org/images/stories/pdf/index2019.pdf.

2. Chris Benner and Kung Feng, "Elon Musk Reflects Silicon Valley's 'Move Fast and Break Things' Culture," *San Francisco Chronicle,* May 15, 2020.

3. John B. Parr, "Growth-Pole Strategies in Regional Economic Planning: A Retrospective View. Part 1. Origins and Advocacy," *Urban Studies* 36, no. 7 (1999): 1195–1215; Casey J. Dawkins, "Regional Development Theory: Conceptual Foundations, Classic Works, and Recent Developments," *Journal of Planning Literature* 18, no. 2 (2003): 140; Ugo Rossi, "Growth Poles, Growth Centers," in *International Encyclopedia of Human Geography,* 2nd ed., ed. A. Kobayashi (Amsterdam: Elsevier, forthcoming).

4. Ottón Solís, "Subsidizing Multinational Corporations: Is That a Development Policy?," Kellogg Institute Working Paper #381, 2011, https://pdfs.semanticscholar.org/7f81/fb21dc7231ec6abe6b2c5b7920b756e6228c.pdf.

5. Lingwen Zheng and Mildred Warner, "Business Incentive Use among U.S. Local Governments: A Story of Accountability and Policy Learning," *Economic Development Quarterly* 24, no. 4 (2010): 325–36.

6. Margaret A. Dewar, "Why State and Local Economic Development Programs Cause So Little Economic Development," *Economic Development Quarterly* 12, no. 1 (1998): 68–87.

7. David Streitfeld, "Was Amazon's Headquarters Contest a Bait-and-Switch? Critics Say Yes," *New York Times,* November 6, 2018, www.nytimes.com/2018/11/06/technology/amazon-hq2-long-island-city-virginia.html.

8. Andres Gunder Frank, *Capitalism and Underdevelopment in Latin America: Historical Studies of Chile and Brazil* (New York: Monthly Review Press, 1966).

9. See, for example, Counter Balance, *How Infrastructure Is Shaping the World: A Critical Introduction to Infrastructure Mega-corridors,* December 2017, www.thecornerhouse.org.uk/sites/thecornerhouse.org.uk/files/Mega%20Corridors.pdf.

10. B. Flyvbjerg, N. Bruzelius, and W. Rothengatter, *Megaprojects and Risk: An Anatomy of Ambition* (Cambridge: Cambridge University Press, 2003); Markku Lehtonen, "Evaluating Megaprojects: From the 'Iron Triangle' to Network Mapping," *Evaluation* 20, no. 3 (2014): 278–95.

11. Andrew T. Simpson, "Health and Renaissance: Academic Medicine and the Remaking of Modern Pittsburgh," *Journal of Urban History* 41, no. 1 (2016): 19–27.

12. US Environmental Protection Agency, *Superfund FY 2019 Annual Accomplishments Report* (Washington, DC, 2019), https://semspub.epa.gov /work/HQ/100002479.pdf.

13. See Fairtrade International's website at www.fairtrade.net.

14. Jody Heymann, Aleta Sprague, and Amy Raub, *Advancing Equality: How Constitutional Rights Can Make a Difference Worldwide* (Berkeley: University of California Press, 2020), 22–23.

15. Heyman, Sprague, and Raub, *Advancing Equality,* 49.

16. United Nations, "Universal Declaration of Human Rights," 1948, www .un.org/en/universal-declaration-human-rights/index.html.

17. David Welsh and J. E. Spence, *Ending Apartheid* (Harlow, UK: Longman/Pearson, 2011).

18. Amartya Sen, *Development as Freedom* (New York: Oxford University Press, 1999).

19. Larry Elliott, "London Economy Subsidises Rest of UK, ONS Figures Show," *The Guardian,* May 23, 2017.

20. Scott Brenton, "The Price of Federation: Comparing Fiscal Equalization in Australia, Canada, Germany, and Switzerland," *Regional and Federal Studies* 30, no. 1 (2019), table 1.

21. Brenton, "Price of Federation," 93–111.

22. HUD Exchange, "Community Development Block Grant Entitlement Program," accessed 2020, www.hudexchange.info/programs/cdbg/cdbg-reports -program-data-and-income-limits/.

23. Katharine Bradbury and Bo Zhao, "Measuring Non-school Fiscal Disparities among Municipalities," *National Tax Journal* 62, no. 1 (2009): 25–62.

24. EdBuild, *$23 Billion,* February 2019, https://edbuild.org/content/23 -billion/full-report.pdf.

25. Urban Institute, "How Do School Funding Formulas Work?," November 29, 2017, https://apps.urban.org/features/funding-formulas/.

26. Karl Marx, *Critique of the Gotha Program* (1875; repr., New York: Occultus Books, 2018).

27. Ann Markusen, Peter Hall, Scott Campbell, and Sabina Deitrick, *The Rise of the Gunbelt: The Military Remapping of Industrial America* (New York: Oxford University Press, 1991).

28. See, for example, P. Aggarwal and R. Aggarwal, "Examining Perspectives and Dimensions of Clean Development Mechanism: A Critical Assessment vis-à-vis Developing and Least Developed Countries," *International Journal of Law and Management* 59, no. 1 (2017): 82–101, https://doi.org/10.1108 /IJLMA-09–2015–0050.

29. On the latter, see Taxpayers for Common Sense, "Coal: A Long History of Subsidies," June 11, 2009, www.taxpayer.net/energy-natural-resources/coal -a-long-history-of-subsidies/.

30. Carlo Trigilia, "Why the Italian Mezzogiorno Did Not Achieve a Sustainable Growth: Social Capital and Political Constraints," *Cambio* 2, no. 4 (2019): 146.

31. Richard Florida, "A Guide to Successful Place-Based Economic Policies," Bloomberg CityLab, March 26, 2019.

32. David C. Korten, "Third Generation NGO Strategies: A Key to People-Centered Development," *World Development* 15 Suppl. (1987): 145–59.

33. See Global Brigades home page, accessed March 21, 2021, www.globalbrigades.org.

34. See Habitat for Humanity, "Habitat's History," accessed March 22, 2021, www.habitat.org/about/history.

35. Stats from 2017. See Catholic Volunteer Network, "Who We Are: Impact," accessed March 22, 2021, https://catholicvolunteernetwork.org/who-we-are/impact/.

36. Ta-Nehisi Coates, "The Case for Reparations," in *We Were Eight Years in Power: An American Tragedy* (New York: One World/Random House, 2017).

37. Thai Jones, "Slavery Reparations Seem Impossible. In Many Places, They're Already Happening," *Washington Post,* January 31, 2020.

38. Neil Vigdor, "North Carolina City Approves Reparations for Black Residents," *New York Times,* July 16, 2020.

39. Dean Baker, "Making a Workable Tobin Tax: A Response to Helmut Reisen," Global Policy Forum, May 2002, https://archive.globalpolicy.org/socecon/glotax/currtax/2002/5tobincritique.htm.

40. Jim Zarroli, "How Bernie Sanders' Wall Street Tax Would Work," National Public Radio, February 12, 2016.

6. HOW CAN WE GET WHERE WE NEED TO GO MORE SUSTAINABLY?

1. Mario Koran, "Oakland to Open Up 74 Miles of City Streets to Pedestrians and Cyclists," *The Guardian,* April 10, 2020.

2. Natalie Colarossi, "Photos Show How Cities Have Closed Streets to Cars So People Have Enough Space to Get Outside during the Pandemic," *Insider,* May 15, 2020.

3. Johnny Diaz, "Cities Close Streets to Cars, Opening Space for Social Distancing," *New York Times,* April 11, 2020.

4. Union of Concerned Scientists, "Cars and Global Warming," 2014, www.ucsusa.org/clean-vehicles/car-emissions-and-global-warming#.XGSFI62ZOgA; Union of Concerned Scientists, "Vehicles, Air Pollution, and Human Health," 2014, www.ucsusa.org/resources/vehicles-air-pollution-human-health.

5. Yuyu Chen, Avraham Ebenstein, Michael Greenstone, and Hongbin Li, "Evidence on the Impact of Sustained Exposure to Air Pollution on Life Expectancy from China's Huai Rivers Policy," *Proceedings of the National Academy of Sciences* 110, no. 32 (2013): 12936–41.

6. Michael Greenhouse and Claire Quin Fan, "Introducing the Air Quality Life Index," Energy Policy Institute, University of Chicago, November 2018, https://aqli.epic.uchicago.edu/wp-content/uploads/2018/11/AQLI_12-facts-update_.pdf.

7. Richard Florida, "How Cars Divide America," Bloomberg CityLab, July 19, 2018.

8. Tamra Johnson, "Americans Spend an Average of 17,600 Minutes Driving Each Year," AAA Newsroom, September 8, 2016, https://newsroom.aaa.com/2016/09/americans-spend-average-17600-minutes-driving-year/.

9. Daniel Sperling, *Three Revolutions: Steering Automated, Shared, and Electric Vehicles to a Better Future* (Washington, DC: Island Press, 2018).

10. Venkat Sumantran, Charles Fine, and David Gonsalvez, *Faster, Smarter, Greener: The Future of the Car and Urban Mobility* (Cambridge, MA: MIT Press, 2017), 292.

11. City of Vancouver, *Active Transportation Promotion and Enabling Plan: Background Report,* accessed March 21, 2021, https://vancouver.ca/files/cov/active-transportation-promotion-and-enabling-full-plan.pdf.

12. Philip Reed and Nicole Arata, "What Is the Total Cost of Owning a Car?," June 28, 2019, www.nerdwallet.com/article/loans/auto-loans/total-cost-owning-car.

13. Tony Dutzik, Jeff Inglis, and Phineas Baxandall, *Millennials in Motion: Changing Travel Habits of Young Americans and the Implications for Public Policy* (Washington, DC: US PIRG Education Fund, 2014).

14. Venu Garikapati, Ram M. Pendyala, Eric A. Morris, Patricia L. Mokhtarian, and Noreen McDonald, "Activity Patterns, Time Use, and Travel of Millennials: A Generation in Transition?," *Transport Reviews* 3, no. 5 (2016): 558–84.

15. Urban Land Institute, "ULI Development Case Studies: Stapleton," December 2015, https://casestudies.uli.org/wp-content/uploads/2015/12/C034004.pdf.

16. Erik Kirschbaum, "Copenhagen Has Taken Bicycle Commuting to a Whole New Level," *Los Angeles Times,* August 8, 2019.

17. Christina Goldbaum, "Is the Subway Risky? It May Be Safer Than You Think," *New York Times,* August 2, 2020.

18. L.A. Metro, "Facts at a Glance," last updated June 24, 2018, www.metro.net/news/facts-glance/.

19. See, for example, Viknesh Vijayenthiran, "Toyota Accelerates Target for EV with Solid-State Battery to 2020," *Motor Authority,* June 10, 2019.

20. Sonia Sodha, "A Radical Way to Cut Emissions—Ration Everyone's Flights," *The Guardian,* May 9, 2018.

21. European Environment Agency, "Transport Fuel Prices and Taxes in Europe," accessed 2018, www.eea.europa.eu/data-and-maps/indicators/fuel-prices-and-taxes/assessment-2.

22. Steve Mufson and James McAuley, "France's Protesters Are Part of a Global Backlash against Climate-Change Taxes," *Washington Post,* December 4, 2018.

23. Donald Shoup, *The High Cost of Free Parking,* updated ed. (New York: Routledge, 2017).

24. David Gutman, "The Not-So-Secret Trick to Cutting Solo Car Commutes: Charge for Parking by the Day," *Seattle Times,* August 10, 2017.

25. Nichole Badstuber, "London's Congestion Charge Is Showing Its Age," Bloomberg CityLab, April 11, 2018, www.citylab.com/transportation/2018/04/londons-congestion-charge-needs-updating/557699/.

26. Transport for London, "Why We Need the ULEZ," accessed 2019, https://tfl.gov.uk/modes/driving/ultra-low-emission-zone/why-we-need-ulez?intcmp=52224; Mayor of London, "Latest Data Shows Two Million Londoners Living with Illegal Toxic Air," April 1, 2019, www.london.gov.uk/press-releases/mayoral/two-million-londoners-live-with-illegal-toxic-air.

27. Transport for London, "Scrappage Scheme," accessed 2019, https://tfl.gov.uk/modes/driving/ultra-low-emission-zone/scrappage-scheme; Mayor of London, "Mayor Doubles 'Scrap for Cash' Polluting Vehicle Fund," press release, February 4, 2019, www.london.gov.uk/press-releases/mayoral/mayor-doubles-scrap-for-cash-dirty-vehicle-fund.

28. Laura Bliss, "The Automotive Liberation of Paris," Bloomberg CityLab, January 19, 2018, www.citylab.com/transportation/2018/01/the-automotive-liberation-of-paris/550718/.

29. Feargus O'Sullivan, "Cities around Paris Strike a Rare Agreement to Ban Diesel Cars," Bloomberg CityLab, November 14, 2018, www.citylab.com/transportation/2018/11/ile-de-france-metropole-paris-diesel-car-ban/575710/.

30. "In Beijing, You have to Win a License Lottery to Buy a New Car," Bloomberg News, February 27, 2019.

31. SEPTA, "Children's Fares," accessed 2020, www.septa.com/fares/discount/children.html.

32. Terry Nguyen, "Kansas City Is Making Its Bus System Fare-Free. Will Other Cities Do the Same?," Vox, December 18, 2019.

33. Emily Badger, "Does Your City Need a Transit Riders Union?," Bloomberg CityLab, October 29, 2012, www.citylab.com/solutions/2012/10/does-your-city-need-transit-riders-union/3722/.

34. Labor Community Strategy Center, "Projects: Bus Riders Union," accessed 2018, https://thestrategycenter.org/projects/bus-riders-union/.

7. HOW DO WE MANAGE LAND MORE SUSTAINABLY?

1. Gregory Clark and Anthony Clark, "Common Rights to Land in England, 1475–1839," *Journal of Economic History* 61, no. 4 (2001): 1009–36.

2. Shengji Pei, Guoxue Zhang, and Huyin Huai, "Application of Traditional Knowledge in Forest Management: Ethnobotanical Indicators of Sustainable Forest Use," *Forest Ecology and Management* 257, no. 10 (2009): 2017–21.

3. Elinor Ostrom, "Common-Pool Resources and Institutions: Toward a Revised Theory," in *Handbook of Agricultural Economics,* ed. Bruce L. Gardner and Gordon C. Rausser (Amsterdam: Elsevier, 2002), vol. 2, pt. A, 1315–39; Stanley Crawford, *Mayordomo: Chronicle of an Acequia in Northern New Mexico* (Albuquerque: University of New Mexico Press, 1988).

4. Andro Linklater, *Owning the Earth: The Transforming History of Land Ownership* (New York: Bloomsbury, 2013).

5. On "production home-building" companies, see Marc A. Weiss, *The Rise of the Community Builders: The American Real Estate Industry and Urban Land Planning* (New York: Columbia University Press, 1987).

6. John R. Logan and Harvey L. Molotch, *Urban Fortunes: The Political Economy of Place* (Berkeley: University of California Press, 1987).

7. An excellent summary of such rights by nation is available at https://en.wikipedia.org/wiki/Freedom_to_roam.

8. Jefferey Sellers, "Urbanization and the Social Origins of National Policies toward Sprawl," in *Urban Sprawl in Western Europe and the United States,* ed. Harry W. Richardson and Chang-Hee Christine Bae (New York: Ashgate, 2004), chap. 12.

9. Ann L. Strong, *Land Banking: European Reality, American Prospect* (Baltimore: Johns Hopkins University Press, 1979).

10. Aldo Leopold, *A Sand County Almanac* (New York: Oxford University Press, 1949).

11. Mihnea Tanasescu, "Rivers Get Human Rights: They Can Sue to Protect Themselves," *The Conversation,* June 19, 2017, https://theconversation.com/when-a-river-is-a-person-from-ecuador-to-new-zealand-nature-gets-its-day-in-court-79278.

12. Christina Rosan, *Governing the Fragmented Metropolis: Planning for Regional Sustainability* (Philadelphia: University of Pennsylvania Press, 2016).

13. Timothy Beatley, "Dutch Green Planning More Reality Than Fiction," *Journal of the American Planning Association* 67, no. 1 (2001): 98–100.

14. Ann Forsyth, "The British New Towns: Lessons for the World from the New-Town Experiment," *Town Planning Review* 90, no. 3 (2019): 239–46.

15. David N. Bengston and Yeo-Chang Youn. "Urban Containment Policies and the Protection of Natural Areas: The Case of Seoul's Greenbelt," *Ecology and Society* 11, no. 1 (2006), www.ecologyandsociety.org/vol11/iss1/art3/.

16. Richard D. Knowles, "Transit-Oriented Development in Copenhagen, Denmark: From the Finger Plan to Orestad," *Journal of Transport Geography* 22 (2012): 251–61.

17. Metro Vancouver, *Metro Vancouver 2040: Shaping Our Future,* July 29, 2011, www.metrovancouver.org/services/regional-planning/metro-vancouver-2040/about-metro-2040/; Thomas A. Hutton, "Thinking Metropolis: From the 'Livable Region' to the 'Sustainable Metropolis' in Vancouver," *International Planning Studies* 16, no. 3 (2011): 237–55.

18. Christina D. Rosan, *Governing the Fragmented Metropolis* (Philadelphia: University of Pennsylvania Press, 2016); Stephen M. Wheeler, "Planning for Metropolitan Sustainability," *Journal of Planning Education and Research* 20 (2000): 133–45.

19. Christopher Alexander, Sara Ishikawa, and Murray Silverstein, *A Pattern Language: Towns, Buildings, Construction* (New York: Oxford University Press, 1977).

20. State of Oregon, *Oregon Conservation Strategy* (Salem: Oregon Department of Fish and Wildlife, 2016), www.oregonconservationstrategy.org.

21. City of Portland, "Ezones Map Correction Project," accessed March 23, 2021, www.portland.gov/bps/ezones/about-ezones-map-correction-project.

22. See, for example, Mike Jenks, Elizabeth Burton, and Katie Williams, eds., *The Compact City: A Sustainable Urban Form?* (London: E. and F.N. Spon, 1996).

23. Stephan Schmidt and Ralph Buehler, "The Planning Process in the US and Germany: A Comparative Analysis," *International Planning Studies* 12, no. 1 (2007): 55–75.

24. Rolf Moeckel and Rebecca Lewis, "Two Decades of Smart Growth in Maryland (U.S.A.): Impact Assessment and Future Directions of a National Leader," *Urban Planning and Transport Research* 5, no. 1 (2017): 22–37.

25. See, for example, Richard T. T. Forman, *Urban Ecology: Science of Cities* (Cambridge: Cambridge University Press, 2015); Monica G. Turner and Robert H. Gardner, *Landscape Ecology in Theory and Practice: Pattern and Process,* 2nd ed. (2000; repr., New York: Springer, 2015).

26. California Department of Water Resources, "Urban Streams Restoration Program," accessed March 23, 2021, https://water.ca.gov/urbanstreams/.

27. Andy Kiersz, "The 20 Biggest Landowners in America," Business Insider, April 16, 2019.

28. Guy Shrubsole, "Who Owns the Country? The Secretive Companies Hoarding England's Land," *The Guardian,* April 19, 2019.

29. Shondell, "12 of the Biggest Land Owners on Earth," October 2, 2015, www.therichest.com/rich-list/the-biggest/12-of-the-biggest-land-owners-on -earth/.

30. US Agency for International Development (USAID), "Country Profile: Guatemala," September 2016, www.usaidlandtenure.net/wp-content/uploads /2016/09/USAID_Land_Tenure_Guatemala_Profile_0.pdf.

31. Vaskela Gediminas, "The Land Reform of 1919–1940: Lithuania and the Countries of East and Central Europe," *Lithuanian Historical Studies/ Lithuanian Institute of History* 1 (1996): 116–32.

32. Henry George, *Progress and Poverty: An Inquiry into the Cause of Industrial Depressions and of Increase of Want with Increase of Wealth: The Remedy* (New York: D. Appleton, 1879).

33. United Nations Habitat, *The Vancouver Declaration on Human Settle- ments,* 1976, Guideline 13, https://unhabitat.org/sites/default/files/download -manager-files/The_Vancouver_Declaration_1976.pdf.

34. Jane H. Malme, "Taxes on Land and Buildings: Case Studies of Transi- tional Economies," *Land Lines* (Lincoln Institute of Land Policy) 11, no. 3 (May 1999): 4–5.

35. William J. McCluskey and Riël C. D. Franzsen. "Property Taxes in Met- ropolitan Cities," in *Financing Metropolitan Governments in Developing Countries,* ed. Roy W. Bahl, Johannes F. Linn, and Deborah L. Wetzel (Cam- bridge: Lincoln Institute of Land Policy, 2013), 176–77.

36. Jeff Wuensch, Frank Kelley, and Thomas Hamilton, "Land Value Taxa- tion Views, Concepts and Methods: A Primer," working paper, Lincoln Institute of Land Policy, Cambridge.

37. Suzi Kerr, Andrew Aitken, and Arthur Grimes, "Land Taxes and Revenue Needs as Communities Grow and Decline," working paper, Lincoln Institute of Land Policy, Cambridge.

38. Garrett Hardin, "The Tragedy of the Commons," *Science* 162, no. 3859 (1968): 1343–48.

39. Elinor Ostrom, *Governing the Commons: The Evolution of Institutions for Collective Action* (Cambridge: Cambridge University Press, 1990).

40. Marin Agricultural Land Trust, "MALT Map and List of Protected Properties," accessed 2018, www.malt.org/MALT-map.

41. David Yetman, "Ejidos, Land Sales, and Free Trade in Northwest Mexico: Will Globalization Affect the Commons?," *American Studies* 41, nos. 2/3 (2000): 211–34.

8. HOW DO WE DESIGN GREENER CITIES?

1. Edward O. Wilson, *Biophilia* (Cambridge, MA: Harvard University Press, 1984).

2. Timothy Beatley, *Biophilic Cities: Integrating Nature into Urban Design and Planning* (Washington, DC: Island Press, 2010).

3. We are paraphrasing Kevin Lynch here. See Kevin Lynch, *A Theory of Good City Form* (Cambridge, MA: MIT Press, 1981).

4. Stephen Burgen, "Barcelona's Car-Free 'Superblocks' Could Save Hundreds of Lives," *The Guardian*, September 10, 2019.

5. Galen Cranz, *The Politics of Park Design: A History of Urban Parks in America* (Cambridge, MA: MIT Press, 1982).

6. Brent Staples, "The Death of the Black Utopia," *New York Times,* November 28, 2019.

7. Jean Gottmann, *Megalopolis: The Urbanized Northeastern Seaboard of the United States* (New York: Twentieth Century Fund, 1961).

8. See, for example, Dave Foreman, *Rewilding North America: A Vision for Conservation in the 21st Century* (Washington, DC: Island Press, 2004); George Monbiot, *Feral: Rewilding the Land, the Sea and Human Life* (Chicago: University of Chicago Press, 2014).

9. Nina Lakhani, "Millions of Americans Lack Access to Quality Parks, Report Reveals," *The Guardian,* May 20, 2020.

10. Megan Heckert and Christina D. Rosan, "Developing a Green Infrastructure Equity Index to Promote Equity Planning," *Urban Forestry and Urban Greening* 19 (2016): 263–70.

11. For more information, see the Seattle Park District's "About" page at www.seattle.gov/seattle-park-district/about.

12. For more information, see National Association of Conservation Districts, "About Districts," accessed March 23, 2021, www.nacdnet.org/about-nacd/about-districts/.

13. See, for example, Yolo Habitat Conservancy, "Frequently Asked Questions," accessed March 23, 2021, www.yolohabitatconservancy.org/faq.

14. See the Santa Monica Mountains Conservancy's website at http://smmc.ca.gov.

15. Michael Van Valkenburgh Associates, *Allegheny Riverfront Park* (New York: Princeton Architectural Press, 2005).

16. See, for example, Catharina Sack, "Landscape Architecture and Novel Ecosystems: Ecological Restoration in an Expanded Field," *Ecological*

Processes 2, no. 35 (2013); Margaret M. Carreiro, Yong-Chang Song, and Jian-guo Wu, eds., *Ecology, Planning, and Management of Urban Forests: International Perspectives* (New York: Springer, 2008).

17. Aga Khan Foundation, "Aga Khan Creates New 30-Hectare Park in Historic Cairo," February 2005, www.akdn.org/press-release/aga-khan-creates-new-30-hectare-park-historic-cairo-media-advisory.

18. Galen Cranz and Michael Boland, "Defining the Sustainable Park: A Fifth Model for Urban Parks," *Landscape Journal* 23, no. 2 (2004): 102–20.

19. See Rijkswaterstaat's website at www.ruimtevoorderivier.nl/english/.

20. ClimateWire, "How the Dutch Make 'Room for the River' by Redesigning Cities," *Scientific American,* January 20, 2012, www.scientificamerican.com/article/how-the-dutch-make-room-for-the-river/.

21. Jeroen Rijke, Sebastiaan van Herk, Chris Zevenbergen, and Richard Ashley, "Room for the River: Delivering Integrated River Basin Management in the Netherlands," *International Journal of River Basin Management* 10, no. 4 (2012): 1814–2060.

22. See, for example, Timothy Beatley, *Blue Urbanism: Exploring Connections between Cities and Oceans* (Washington, DC: Island Press, 2014).

23. Portland Bureau of Transportation, "Livable Streets: A Strategy for Portland," accessed March 23, 2021, www.portlandoregon.gov/transportation/71710.

24. Nancy W. Stauffer, "Transparent Solar Cells: Generating Power from Everyday Surfaces," MIT Energy Initiative press release, June 20, 2013, https://energy.mit.edu/news/transparent-solar-cells/.

25. John Dialesandro, Noli Brazil, Stephen Wheeler, and Yasr Abunnasr, "Dimensions of Thermal Inequity: Relations between Neighborhood Social Demographics and Urban Heat in the Southwestern U.S.," *International Journal of Environmental Research and Public Health* 18, no. 3 (2021); 941, doi: 10.3390/ijerph18030941.

26. See, for example, Paul D. Tennis, Michael L. Leming, and David J. Akers, *Pervious Concrete Pavements* (Skokie, IL: Portland Cement Association, 2004), http://myscmap.sc.gov/marine/NERR/pdf/PerviousConcrete_pavements.pdf.

27. J. Sansalone, X. Kuang, G. Yink, and V. Ranieri, "Filtration and Clogging of Permeable Pavement Loaded by Urban Drainage," *Water Research* 46, no. 20 (2012): 6763–74.

28. National Wildlife Federation, "At Home: Designing Your Wildlife Garden," accessed March 23, 2021, www.nwf.org/sitecore/content/Home/Garden-for-Wildlife/Create/At-Home.

29. For a history of urban gardening in the US, see Laura Lawson, *The City Bountiful: A Century of Community Gardening in America* (Berkeley: University of California Press, 2005).

30. See, for example, the website of National Black Food & Justice Alliance at www.blackfoodjustice.org.

31. National Conference of State Legislatures, "States Shut Out Light Pollution," May 23, 2016, www.ncsl.org/research/environment-and-natural-resources/states-shut-out-light-pollution.aspx.

9. HOW DO WE REDUCE OUR ECOLOGICAL FOOTPRINTS?

1. Joel Makower, "The Death and Rebirth of *50 Simple Things You Can Do to Save the Earth*," GreenBiz, March 22, 2008.

2. John Javna, Sophie Javna, and Jesse Javna, *50 Simple Things You Can Do to Save the Earth,* rev. ed. (New York: Hyperion Books, 2008), 11.

3. Mathis Wackernagel and William Rees, *Our Ecological Footprint* (Gabriola, BC: New Society, 1996).

4. Christopher M. Jones and Daniel M. Kammen, "Quantifying Carbon Footprint Reduction Opportunities for U.S. Households and Communities," *Environmental Science and Technology* 45 (2011): 4088–95.

5. Mario Herrero, Petr Havlík, Hugo Valin, An Notenbaert, Mariana C. Rufino, Philip K. Thornton; Michael Blümmel, et al., "Biomass Use, Production, Feed Efficiencies, and Greenhouse Gas Emissions from Global Livestock Systems," *Proceedings of the National Academy of Sciences* 110, no. 52 (2013): 20888–93.

6. Petr Havlik, Hugo Valin, Mario Herrero, Michael Obersteiner, Erwin Schmid, Mariana C Rufino, Aline Mosnier, et al., "Climate Change Mitigation through Livestock System Transitions," *Proceedings of the National Academy of Sciences* 111, no. 10 (2014): 3709–14.

7. Matthew Nitch Smith, "The Number of Cars Worldwide Is Set to Double by 2040. World Economic Forum," April 22, 2016, www.weforum.org /agenda/2016/04/the-number-of-cars-worldwide-is-set-to-double-by-2040.

8. Michael I. Goran, Stanley J. Ulijaszek, and Emily E. Ventura, "High Fructose Corn Syrup and Diabetes Prevalence: A Global Perspective," *Global Public Health* 8, no. 1 (2014): 55–64.

9. A. Wolk, "Potential Health Hazards of Eating Red Meat," *Journal of Internal Medicine* 281, no. 2 (2016): 106–22.

10. Andy Murdock, "The Diet That Helps Climate Change," Vox, December 12, 2017.

11. Amanda Coetzee, "Food in the Nude: NZ Ditching Plastic Packaging of Fruits and Vegetables," *Mpumalanga News,* January 22, 2019.

12. Adele E. Clarke and Donna Haraway, *Make Kin Not Population* (Chicago: Prickly Paradigm Press, 2018).

13. Albert Bandura, *Social Foundations of Thought and Action: A Social Cognitive Theory* (Englewood Cliffs, NJ: Prentice Hall, 1986).

14. Icek Ajzen, "The Theory of Planned Behavior," *Organizational Behavior and Human Decision Processes* 50, no. 2 (1991): 179–211.

15. James O. Prochaska and Wayne F. Velicer, "The Transtheoretical Model of Behavior Change," *American Journal of Health Promotion* 12, no. 1 (1997): 38–48.

16. B. J. Fogg, "A Behavior Model for Persuasive Design," in *Proceedings of the 4th International Conference on Persuasive Technology* (New York: ACM, 2009), 40.

17. B. J. Fogg, "The New Rules of Persuasion," *RSA Journal* 155, no. 5538 (2009): 24–29.

18. Doug McKenzie-Mohr, *Fostering Sustainable Behavior: An Introduction to Community-Based Social Marketing* (Gabriola Island, BC: New Society, 2006).

19. Ingrid Robeyns, "What, if Anything, Is Wrong with Extreme Wealth?," *Journal of Human Development and Capabilities* 20, no. 3 (2019): 251–66.

20. Wessel P. Visser, "A Perfect Storm: The Ramifications of Cape Town's Drought Crisis," *Journal for Transdisciplinary Research in Southern Africa* 14, no. 1 (2018): 1–10.

10. HOW CAN CITIES BETTER SUPPORT HUMAN DEVELOPMENT?

1. Peter Hall, *Cities of Tomorrow: An Intellectual History of Urban Planning and Design since 1880* (Chichester, UK: Wiley, 2014).

2. Steven Pinker, *The Better Angels of Our Nature: Why Violence Has Declined* (New York: Viking, 2011); Jared Diamond, *The World until Yesterday: What Can We Learn from Traditional Societies* (New York: Viking, 2012).

3. See 2021 and earlier reports at Sustainable Development Solutions Network, "World Happiness Report," https://worldhappiness.report.

4. Lewis Mumford, *The City in History* (New York: Harcourt, 1961).

5. Jane Jacobs, *The Death and Life of Great American Cities* (New York: Vintage, 1961).

6. Kevin Lynch, *A Theory of Good City Form* (Cambridge, MA: MIT Press, 1981); Christopher Alexander, Sara Ishikawa, and Murray Silverstein, *A Pattern Language: Towns, Buildings, Construction* (New York: Oxford University Press, 1977).

7. Randolph T. Hester, *Design for Ecological Democracy* (Cambridge, MA: MIT Press, 2006).

8. Nan Ellin, *Post-modern Urbanism* (New York: Princeton Architectural Press, 1996).

9. Nan Ellin, *Good Urbanism: Six Steps to Creating Prosperous Places* (Washington, DC: Island Press, 2013).

10. Emily Talen, *Urban Design for Planners: Tools, Techniques, and Strategies*, 2nd ed. (Los Angeles: Planetizen Press, 2018).

11. William H. Whyte, *The Social Life of Small Urban Spaces* (Washington, DC: Conservation Foundation, 1980).

12. Steven Kaplan, "The Restorative Benefits of Nature: Toward an Integrative Framework," *Journal of Environmental Psychology* 15 (1995): 169–82.

13. Emily Hannum and Claudia Buchmann, "Global Educational Expansion and Socio-Economic Development: An Assessment of Findings from the Social Sciences," *World Development* 33, no. 3 (2005): 333–54.

14. Edward L. Glaeser, Giacomo A. M. Ponzetto, and Andrei Schleifer, "Why Does Democracy Need Education?," *Journal of Economic Growth* 12 (2007): 77–99.

15. W. Steven Barnett and Ellen C. Frede, "Long-Term Effects of a System of High-Quality Universal Preschool Education in the United States," in *Childcare, Early Education and Social Inequality: An International Perspective,* ed. Hans-Peter Blossfeld, Nevena Kulic, Jan Skopek, and Moris Triventi (Cheltenham, UK: Edward Elgar, 2017), 152–72.

16. Gus Wezerek, "The Wealthy Can't Stop Not Paying Their Taxes," *New York Times,* July 15, 2020.

17. Leslie Kern, "How to Rebuild Cities for Caregiving," Bloomberg City-Lab, July 9, 2020, www.bloomberg.com/news/articles/2020–07–09/a-feminist-vision-for-supporting-urban-caregivers.

18. Nicholas Kristof, "McDonald's Workers in Denmark Pity Us," *New York Times,* May 8, 2020.

19. See, for example, David de la Pena, Diane Jones Allen, Randolph T. Hester, Jeffrey Hou, Laura J. Lawson, and Marcia J. McNally, eds., *Design as Democracy: Techniques for Collective Creativity* (Washington, DC: Island Press, 2018).

11. HOW MIGHT WE HAVE MORE FUNCTIONAL DEMOCRACY?

1. Anthony Alexander, *Britain's New Towns: Garden Cities to Sustainable Communities* (London: Routledge, 2009).

2. Sherry Arnstein, "A Ladder of Citizen Participation," *Journal of the American Planning Association* 35, no. 4 (1969): 216–24.

3. Paul Davidoff, "Advocacy and Pluralism in Planning," *Journal of the American Planning Association* 31, no. 4 (1964): 331–38.

4. Tony Bovaird, "Beyond Engagement and Participation: User and Community Coproduction of Public Services," *Public Administration Review* 67, no. 5 (2007): 846–60.

5. Katharine Coit, "Participation, Social Movements and Social Change," *Cities,* November 1984, 584–91; Enrique Ortecho, S. Sabagh de Pipa, and Maria Cristina Bosio de Ortecho, "Participation Experiences," *Cities,* November 1984, 580–84.

6. Christine Cheyne, "Changing Urban Governance in New Zealand: Public Participation and Democratic Legitimacy in Local Authority Planning and Decision-Making, 1989–2014," *Urban Policy and Research* 33, no. 4 (2015): 416–32.

7. Thomas Frank, *What's the Matter with Kansas? How Conservatives Won the Heart of America* (New York: Henry Holt, 2005).

8. See, for example, Paul Matzko, "Talk Radio Is Turning Millions of Americans into Conservatives," *New York Times,* October 9, 2020.

9. Craig Holman and William Luneburg, "Lobbying and Transparency: A Comparative Analysis of Regulatory Reform," *Interest Groups and Advocacy* 1, no. 1 (2012): 75–104.

10. Tacey Rychter, "Australia Tells America: Here's How to Fix Your Voting System," *New York Times,* October 22, 2018.

11. Velmyndigheten (Swedish Government agency), "How to Vote," accessed 2018, www.val.se/download/18.5af422c516299cedac1fb5/1538729830824/Att%20rösta%20(engelska).pdf.

12. Cristina Costantini, "3 Countries Where It's Easier to Vote Than the United States," ABC News, November 2, 2012.

13. Paul Waldman, "How Our Campaign Finance System Compares to Other Countries," *American Prospect,* April 4, 2014.

14. Waldman, "How Our Campaign Finance System Compares."

15. Michael P. McCauley, *NPR: The Trials and Triumphs of National Public Radio* (New York: Columbia University Press, 2005).

16. Victor Pickard, *America's Battle for Media Democracy: The Triumph of Corporate Libertarianism and the Future of Media Reform* (New York: Cambridge University Press, 2013).

17. Ashley Lutz, "These Six Companies Control 90% of the Media in America," Business Insider, December 6, 2018.

18. Jack Morse, "Facebook Illegally Maintains a Monopoly and a Breakup Is on the Table, Says FTC Lawsuit," MSNBC.com, December 10, 2020, www .msn.com/en-us/news/technology/facebook-illegally-maintains-a-monopoly-and -a-breakup-is-on-the-table-says-ftc-lawsuit/ar-BB1bN2Sr.

19. Laura Johnson and Paul Morris, "Towards a Framework for critical Citizenship Education," *Curriculum Journal* 21, no. 1 (2010): 77–96.

20. Fernando Reimers, "Civic Education When Democracy Is in Flux: The Impact of Empirical Research on Policy and Practice in Latin America," *Citizen Teaching and Learning* 3, no. 2 (2007): 5–21.

21. Benjamin Barber, *Strong Democracy: Participatory Politics for a New Age* (1984; repr., Berkeley: University of California Press, 2003).

22. John S. Dryzek, *Deliberative Democracy and Beyond: Liberals, Critics, Contestations* (Oxford: Oxford University Press, 2002).

23. Robert D. Putnam, *Making Democracy Work: Civic Traditions in Modern Italy* (Princeton, NJ: Princeton University Press, 1994).

24. PlannersWeb, "Holding Effective Public Meetings," October 16, 2014, http://plannersweb.com/2014/10/holding-effective-public-meetings/.

25. Center for Advances in Public Engagement, "Public Engagement: A Primer from Public Agenda," June 15, 2015, https://metrocouncil.org /Handbook/Files/Community-Engagement/PublicEngagementPrimer.aspx.

26. See the website of Philadelphia Citizen Planning Institute at https:// citizensplanninginstitute.org.

27. Michael Green, "People's Panel Pitches In to Advise Melbourne City Council Where It Should Spend $5 Billion," *The Age,* December 2, 2014.

28. See, for example, Jennifer Evans-Cowley and Justin Hollander, "The New Generation of Public Participation: Internet-Based Tools," *Planning Practice and Research* 25, no. 3 (2010): 397–408; Reinout Kleinhans, Maarten Van Ham, and Jennifer Evans-Cowley, "Using Social Media and Mobile Technologies to Foster Engagement and Self-Organization in Participatory Urban Planning and Neighborhood Governance," *Planning Practice and Research* 30, no. 3 (2015): 237–47.

29. Sarah Holder, "Can the Blockchain Tame Moscow's Wild Politics?," Bloomberg CityLab, December 22, 2017, www.citylab.com/life/2017/12/can -the-blockchain-tame-moscows-wild-politics/547973/.

30. Maarit Kahila-Tani, Anna Broberg, Marketta Kyttä, and Taylor Tyger, "Let the Citizens Map—Public Participation GIS as a Planning Support System in the Helsinki Master Plan Process," *Planning Practice and Research* 31, no. 2 (2016): 195–214.

31. Michael Touchton and Brian Wampler, "Improving Social Well-Being through New Democratic Institutions," *Comparative Political Studies* 47, no. 10 (2014): 1442–69.

32. Enrique Peruzzotti, "La Plata, Argentina: Multi-channel Participatory Budgeting Case Study," 2011, https://participedia.net/case/1263.

33. Elena Krylova, "Participatory Budgeting in Ukraine: Current Practices and Implications for Their Advancement," World Bank, 2006, http://siteresources .worldbank.org/INTPSIA/Resources/490023–1120841262639/PSIA_Participatory _Budgeting_in_Ukraine_English.pdf.

34. Daniel Fisher, "A Look at New York City's Participatory Budgeting Map," *Digital Communitie,* August 7, 2018, www.govtech.com/dc/A-Look-at -New-York-Citys-Participatory-Budgeting-Map.html.

35. Rosario del Pilar Díaz Garavito, "Programa de Embajadores Peru Agenda 2030: Por el desarrollo sostenible y el mundo que queremos," United Nations Sustainable Development Goals Action Campaign, March 16, 2017, https:// sdgactioncampaign.org/2017/03/16/programa-de-embajadores-peru-agenda -2030-por-el-desarrollo-sostenible-y-el-mundo-que-queremos/.

36. Saransh Sehgal, "Why Vienna Does So Well on Quality-of-Life Rankings," Bloomberg CityLab, August 23, 2016, www.citylab.com/equity/2016/08 /why-vienna-always-does-well-on-quality-of-life-rankings/496997/.

37. Smart City Wien, "Monitoring Report 2017: Smart City Wien Framework Strategy," February 26, 2018, https://smartcity.wien.gv.at/en/monitoring -report-2017/.

12. HOW CAN EACH OF US HELP LEAD THE MOVE TOWARD SUSTAINABLE COMMUNITIES?

1. See, for example, David Harvey, *The Ways of the World* (New York: Oxford University Press, 2016); Manuel Castells, *Networks of Outrage and Hope: Social Movements in the Internet Age,* 2nd ed. (Cambridge: Polity Press, 2015); Doreen Massey, *World City* (Malden, MA: Policy Press, 2007); Mike Davis, *City of Quartz: Excavating the Future in Los Angeles* (New York: Verso, 1990).

2. See, for example, Walter W. Powell and Paul J. DiMaggio, *The New Institutionalism in Organizational Analysis* (1991; repr., Chicago: University of Chicago Press, 2012); Richard W. Scott, *Institutions and Organizations: Ideas, Interests and Identities* (Thousand Oaks, CA: Sage Publications, 2014).

3. Paul Davidoff, "Advocacy and Pluralism in Planning," *Journal of the American Planning Association* 31, no. 4 (1965): 331–38.

4. See, for example, Leoni Sandercock, *Cosmopolis II: Mongrel Cities of the 21st Century* (New York: Continuum, 2003); Jeff Hou, ed., *Insurgent Public Space: Guerrilla Urbanism and the Remaking of Contemporary Cities* (New York: Routledge, 2010).

5. Victoria Beard, "Learning Radical Planning: The Power of Collective Action," *Planning Theory* 2, no. 1 (2003): 13–35.

6. John Friedmann, *Planning in the Public Domain: From Knowledge to Action* (Princeton, NJ: Princeton University Press, 1987).

7. Patsy Healey, *Collaborative Planning: Shaping Places in Fragmented Societies* (Vancouver: University of British Columbia Press, 1997); Judith Innes, "Planning Theory's Emerging Paradigm: Communicative Action and Interactive Practice," *Journal of Planning Education and Research* 14, no. 3 (1995): 183–89.

8. Paolo Freire, *Pedagogy of the Oppressed* (New York: Continuum, 1970).

9. See, for example, Dolores Hayden, *Redesigning the American Dream: Gender, Housing and Family Life*, expanded ed. (1984; repr., New York: Norton, 2002); Leslie Kern, *Feminist City: Claiming Space in a Man-Made World* (New York: Verso, 2020).

10. See, for example, Ta-Nehisi Coates, *We Were Eight Years in Power: An American Tragedy* (New York: One World, 2017).

11. J. W. Kingdon and E. Stano, *Agendas, Alternatives, and Public Policies* (Boston: Little, Brown, 1984).

12. Lawrence Susskind and J. Cruickshank, *Breaking the Impasse: Consensual Approaches to Resolving Public Disputes* (New York: Basic Books, 1987).

13. Barbara Tuchman, *The March of Folly: From Troy to Vietnam* (1984; repr., New York: Random House, 2011).

Index

Founded in 1893,
UNIVERSITY OF CALIFORNIA PRESS
publishes bold, progressive books and journals
on topics in the arts, humanities, social sciences,
and natural sciences—with a focus on social
justice issues—that inspire thought and action
among readers worldwide.

The UC PRESS FOUNDATION
raises funds to uphold the press's vital role
as an independent, nonprofit publisher, and
receives philanthropic support from a wide
range of individuals and institutions—and from
committed readers like you. To learn more, visit
ucpress.edu/supportus.